Handbook of Hydrology for Engineers

Handbook of Hydrology for Engineers

Contributors

Mario Lefebvre, Fatima Bensalma et al.

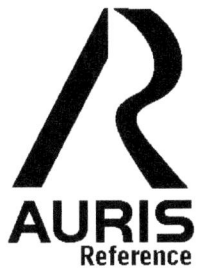

AURIS
Reference

www.aurisreference.com

Handbook of Hydrology for Engineers

Contributors: Mario Lefebvre, Fatima Bensalma et al.

Published by Auris Reference Limited

www.aurisreference.com

United Kingdom

Handbook of Hydrology for Engineers

ISBN: 978-1-78154-830-1

British Library Cataloguing in Publication Data
A CIP record for this book is available from the British Library

Printed in the United Kingdom

Exclusively distributed by CBS Publishers & Distributors Pvt. Ltd.

Sales & Distribution Rights only for India, Pakistan, Bangladesh, Sri Lanka, Nepal and Bhutan.This book is not to be sold outside these territories.

Contents

List of Abbreviations

ANN	Artificial Neural Networks
BMP	Best Management Practice
CLM	Community Land Model
CN	Curve Number
CRWR	Center for Research in Water Resources
DEM	Digital Elevation Model
DHM	Distributed Hydrological Models
DOQQ	Digital Orthophoto Quarter Quad
EFDC	Environmental Fluid Dynamics Code
EKF	Extended Kalman Filter
ESM	Earth System Model
GCM	General Circulation Model
GEEI	Global Energetic Efficiency Indicator
GEV	Generalized Extreme Value
GIS	Geographic Information Science
GIS	Geographic Information Systems
GLEAMS	Groundwater Loading Effects of Agricultural Management Systems
GRACE	Gravity Recovery and Climate Experiment
HRU	Hydrologic Response Unit
ISMN	International Soil Moisture Network
LDHT	Low Dosage Herbicide Treatments
LID	Low Impact Development
LSM	Land Surface Model
MUSLE	Modified Universal Soil Loss Equation
NBR	Normalized Burn Ratio
NLCD	National Land Cover Database
RMSE	Root Mean Square Error
ROTO	Routing Outputs to Outlet
RR	Rainfall-Runoff
RUSLE	Revised Universal Soil Loss Equation
SCS	Soil Conservation Service
SM	Soil Moisture
SUDS	Sustainable Urban Drainage System
SWB	Soil Water Balance
SWMM	Storm Water Management Model
SWRRB	Simulator for Water Resources in Rural Basins
TMDL	Total Maximum Daily Loads
TND	Traditional Neighborhood Development
VIC	Variable Infiltration Capacity
WDM	Watershed Data Management
WEPP	Water Erosion Prediction Project
WMS	Watershed Modeling System
WSUD	Water Sensitive Urban Design

List of Contributors

Mario Lefebvre
Département de Mathématiques et de Génie Industriel, École Polytechnique, C.P. 6079, Succursale Centre-ville, Montréal, QC, Canada H3C 3A7

Fatima Bensalma
Département de Mathématiques et de Génie Industriel, École Polytechnique, C.P. 6079, Succursale Centre-ville, Montréal, QC, Canada H3C 3A7

Prem B. Parajuli
Department of Agricultural and Biological Engineering at Mississippi State University, Mississippi State, USA

Ying Ouyang
USDA-Forest Service Center for Bottomland Hardwoods Research, Mississippi State, USA

Yingying Meng
College of Water Sciences, Beijing Normal University, Beijing 100875, China
Beijing Water Science and Technology Institute, Beijing 100048, China

Huixiao Wang
College of Water Sciences, Beijing Normal University, Beijing 100875, China

Jiangang Chen
Beijing Water Science and Technology Institute, Beijing 100048, China

Shuhan Zhang
Beijing Water Science and Technology Institute, Beijing 100048, China

Dong Jiang
Institute of Geographical Sciences and Natural Resources Research, Chinese Academy of Sciences, Beijing 100101, China

Jianhua Wang
State Key Laboratory of Simulation and Regulation of Water Cycle in River Basin, Department of Water Resources, China Institute of Hydropower & Water Resources Research, Beijing 100038, China

Yaohuan Huang
Institute of Geographical Sciences and Natural Resources Research, Chinese Academy of Sciences, Beijing 100101, China

Kang Zhou
Computer Network Information Center, Chinese Academy of Sciences, Beijing 100190, China

Xiangyi Ding
State Key Laboratory of Simulation and Regulation of Water Cycle in River Basin, Department of Water Resources, China Institute of Hydropower & Water Resources Research, Beijing 100038, China

Jingying Fu
Institute of Geographical Sciences and Natural Resources Research, Chinese Academy of Sciences, Beijing 100101, China

Saeid Eslamian
Isfahan University of Technology, Isfahan, Iran

Kristin L. Gilroy
University of Maryland, USA

Richard H. McCuen
University of Maryland, USA

Ivan N. da Silva
University of São Paulo (USP), São Carlos, SP, Brazil

José Ângelo Cagnon
São Paulo State University (UNESP), Bauru, SP, Brazil

Nilton José Saggioro
University of São Paulo (USP), Bauru, SP, Brazil

Christopher Andrew Day
Department of Geography and Geosciences, University of Louisville, Louisville, USA

Keith Allen Bremer
Department of Geography, Texas State University-San Marcos, San Marcos, USA.

Paul R. Houser
George Mason Univ., Fairfax, VA, USA
Bureau of Reclamation, Washington, DC, USA

Gabriëlle J.M. De Lannoy
Ghent Univ., Ghent, Belgium
NASA Goddard Space Flight Center, Greenbelt, MD, USA

Jeffrey P. Walker
Monash University, Melbourne, Australia

Giuseppe Bombino
Mediterranean University of Reggio Calabria Department of Agro-Forest and Environmental Sciences and Technologies Italy

Vincenzo Tamburino
Mediterranean University of Reggio Calabria Department of Agro-Forest and Environmental Sciences and Technologies Italy

Demetrio Antonio Zema
Mediterranean University of Reggio Calabria Department of Agro-Forest and Environmental Sciences and Technologies Italy

Santo Marcello Zimbone
Mediterranean University of Reggio Calabria Department of Agro-Forest and Environmental Sciences and Technologies Italy

Vitali Diaz Mercado
UNESCO-IHE Institute for Water Education, Delft, Netherlands

Khalidou M. Bâ
Centro Interamericano de Recursos del Agua, Universidad Autónoma del Estado de México, Toluca, México

Emmanuelle Quentin
Instituto Nacional de Investigación en Salud Pública, Quito, Ecuador

Febe Helia Ortiz Madrid
Centro Interamericano de Recursos del Agua, Universidad Autónoma del Estado de México, Toluca, México

Lilly Gama
División Académica de Ciencias Biológicas, Universidad Juárez Autónoma de Tabasco, México

Johanna Springer
Department of Geography, Ludwig-Maximilians-Universitaet Muenchen (LMU), Luisenstr. 37, D-80333 Munich, Germany

Ralf Ludwig
Department of Geography, Ludwig-Maximilians-Universitaet Muenchen (LMU), Luisenstr. 37, D-80333 Munich, Germany

Stefan W. Kienzle
Department of Geography, University of Lethbridge, Alberta Water and Environmental Science Building, 4401 University Drive, Lethbridge, AB T1K-3M4, Canada
Applied Behavioral Ecology and Ecosystems Research Unit, University of South Africa, PO Box 392, Florida, 1710 Pretoria, South Africa

Christian Massari
Research Institute for Geo-Hydrological Protection, National Research Council CNR, Via di Madonna Alta 126, 06128 Perugia, Italy

Luca Brocca
Research Institute for Geo-Hydrological Protection, National Research Council CNR, Via di Madonna Alta 126, 06128 Perugia, Italy

Luca Ciabatta
Research Institute for Geo-Hydrological Protection, National Research Council CNR, Via di Madonna Alta 126, 06128 Perugia, Italy

Tommaso Moramarco
Research Institute for Geo-Hydrological Protection, National Research Council CNR, Via di Madonna Alta 126, 06128 Perugia, Italy

Simone Gabellani
International Centre on Environmental Monitoring (CIMA) Research Foundation, Via A. Magliotto, 2-17100 Savona, Italy

Clement Albergel
European Centre for Medium-Range Weather Forecasts (ECMWF), Shinfield Park, RG2 9AX Reading, UK

Patricia De Rosnay
European Centre for Medium-Range Weather Forecasts (ECMWF), Shinfield Park, RG2 9AX Reading, UK

Silvia Puca
Italian Civil Protection Department, Via Vitorchiano 2, 00189 Rome, Italy

Wolfgang Wagner
Department of Geodesy and Geoinformation, Vienna University of Technology, 1040 Vienna, Austria

Preface

Hydrology is the scientific study of the movement, distribution, and quality of water on Earth and other planets, including the hydrologic cycle, water resources and environmental watershed sustainability. The text Handbook of Hydrology for Engineers covers the environmental aspects of hydrology such as environmental river flow, hydro-fracturing, sustainability in wastewater treatment, water pollution control using low cost natural waste, environmental impacts of nanotechnology, ecological risk on reservoirs, and modeling of wetland systems. The aim of first chapter is to show that we can obtain more accurate short-term forecasts of river flows by making use of filtered renewal processes rather than filtered Poisson processes. Second chapter provides review of several watershed and water quality models and two case studies to evaluate future climate change impact on hydrology using two models. Modelling hydrology of a single bioretention system with HYDRUS-1D has been discussed in third chapter. Fourth chapter presents a review of recent applications of GRACE data in terrestrial hydrology monitoring. The objective of fifth chapter is to describe the statistical techniques for detecting changes in hydrological events. Recurrent neural network based approach for solving groundwater hydrology problems has been presented in sixth chapter. The purpose of seventh chapter is to investigate the potential surface runoff generated from the kind of development in comparison to a traditional neighborhood. Eighth chapter focuses on hydrological data assimilation which is an objective method to estimate the hydrological system states from irregularly distributed observations. Hydrological effects of different soil management practices in Mediterranean areas have been investigated in ninth chapter. Hydrological model to simulate daily flow in a basin with the help of a GIS has been focused in tenth chapter. Eleventh chapter investigates the hydrology of Castle River in the southern Canadian Rocky Mountains. The use of H-SAF soil moisture products for operational hydrology has been discussed in last chapter.

Chapter 1

AN APPLICATION OF FILTERED RENEWAL PROCESSES IN HYDROLOGY

Mario Lefebvre and Fatima Bensalma

Département de Mathématiques et de Génie Industriel, École Polytechnique, C.P. 6079, Succursale Centre-ville, Montréal, QC, Canada H3C 3A7

ABSTRACT

Filtered renewal processes are used to forecast daily river flows. For these processes, contrary to filtered Poisson processes, the time between consecutive events is not necessarily exponentially distributed, which is more realistic. The model is applied to obtain one- and two-day-ahead forecasts of the flows of the Delaware and Hudson Rivers, both located in the United States. Better results are obtained than with filtered Poisson processes, which are often used to model river flows.

INTRODUCTION

Let $\{(t), t \geq 0\}$ be a Poisson process with rate λ. A filtered Poisson (sometimes called shot noise) process is a continuous-time stochastic process $\{(t), t \geq 0\}$ defined by

$$X(t) = \sum_{n=1}^{N(t)} w(t, \tau_n, Y_n) \quad (X(t) = 0 \text{ if } N(t) = 0),$$

(1)

in which the random variables $\tau_1, \tau_2, \ldots$ denote the arrival times of the events of the Poisson process and $Y_1, Y_2, \ldots$ are assumed to be independent and identically distributed random variables that are also independent of $\{N(t), t \geq 0\}$. The function $(\cdot, \cdot, \cdot)$ is called the response function.

In many applications, the response function is chosen of the form

$$w(t, \tau_n, Y_n) = Y_n e^{-(t-\tau_n)/c},$$

(2)

where c is a parameter that must be estimated. It then gives the value at time t of an event of magnitude Y_n of the Poisson process that occurred at time

τ_n. Moreover, the random variables Y_1, Y_2,... are generally assumed to be exponentially distributed with parameter μ. With the above response function, the filtered Poisson process behaves as in Figure 1. Actually, this behavior depends on the form of the response function, but not on the distribution of the time between the events. Therefore, the same behavior would be observed in the case when $\{(t), t \geq 0\}$ is a renewal process. Remember that a Poisson process is a particular renewal process.

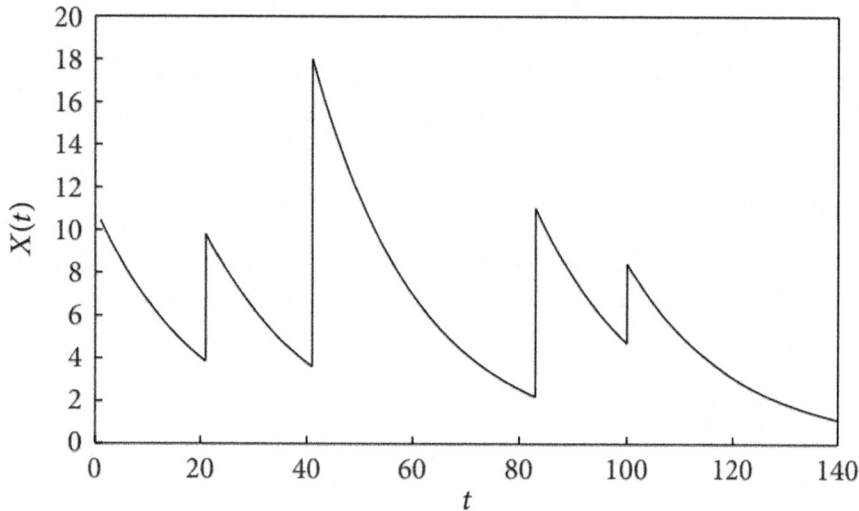

Figure 1: Example of a trajectory of a filtered Poisson process defined by (1) and (2).

To model the daily flows of rivers given their most recent observed values, conceptual and physical models based on the filtered Poisson process have been widely used successfully for many decades; see, for example, Weiss [1], Kelman [2], Koch [3], and Konecny [4]. Filtered Poisson processes are still being used to model various phenomena in civil engineering; see Yin et al. [5] and Miyamoto et al. [6].

Now, especially in hydrological applications, the form of the response function in (2) is taken for granted rather than being justified by the observations of the variable of interest. Actually, it is generally simply an assumption made to obtain a mathematically tractable model.

Similarly, the actual distribution of the Y_n's is not investigated. However, if one is only interested in forecasting the value of $(t+\delta)$, based on the values of the process up to time t, this does not really cause a problem because the forecast is normally based on the mean of the Y_n's, and the distribution itself is not needed.

Finally, and even more importantly, the assumption that $\{(t), t \geq 0\}$ is indeed a Poisson process is not tested. Again, this is a simplifying assumption, because one can then make use of the nice properties of the Poisson process, notably the fact that it has independent and stationary increments.

Next, notice that, with the response function in (2), the effect on the value of the process of an event that occurred at time τ_n is maximum at that time instant and $\{X(t), t \geq 0\}$ has discontinuities at the arrival times of the events of the Poisson process. In the case of the application in hydrology that consists in modeling the flow of rivers, if one looks at real hydrographs, one sees that there is a more or less extended period of time during which the flow increases from a minimum to a maximum. Even if river flows data are generally available on a daily basis, one does not observe very sudden increases of the flow, followed by an exponential decrease. Rather, it takes on average two to three days for the flow to reach a maximum, and the decrease is more or less rapid.

To obtain a more realistic model for the variations of river flows, by taking into account the periods of flow increase, Lefebvre and Guilbault [7] used the following response function:

$$w\left(t, \tau_n, Y_n\right) = Y_n\left(t - \tau_n\right)^d e^{-(t-\tau_n)/c},$$

(3)

where d is a positive parameter. With this response function, the river flow begins to increase immediately after time τ_n, but the maximum is reached after dc time units (i.e., generally days) and then it starts to decrease. Their work was improved by the authors, who provided a method to estimate the parameter d. They found that, for the applications that they considered, d was in the interval $(0,1)$.

With a Poisson process being a continuous-time Markov chain, the time that the process spends in a given state is exponentially distributed, which is a strictly decreasing density function. Let T_1, T_2,... be the times between two successive events, that is, the interarrival times. In the case of river flows, the interarrival time is generally a random variable with a density function that is increasing at first. Hence, the assumption that $\{(t), t \geq 0\}$ is a Poisson process is at least doubtful.

In Lefebvre [8], the author tested the hypothesis that the T_n's are exponentially distributed. He found that, for the Delaware River, it was not accurate. Instead, it was shown that both a gamma and a Rayleigh distribution fitted the data much better. The objective was to forecast the peak flows of the Delaware River, which is a difficult task due to the very weak correlation between the successive peaks. To try to forecast these peak flows, the author used a filtered renewal process as a model for the river flow. An advantage in using a Rayleigh distribution for the interarrival time is that the filtered renewal process can be transformed into a filtered Poisson process by working on a different time scale.

Poisson and renewal processes have been used in hydrology to model low flows (see, for instance, Loaiciga and Leipnik [9] and Yagouti et al. [10]) as well as various physical phenomena such as traffic noise (see Marcus [11]). Andersen [12, 13] gave new theoretical results on filtered renewal processes (see also the references therein).

The aim of the present paper is to show that we can obtain more accurate short-term forecasts of river flows by making use of filtered renewal processes rather than filtered Poisson processes. To do so, we will first need to develop a formula to forecast the flow at time $(t + \delta)$, given the history of the process up to time t. Because we cannot appeal to the memoryless property of the exponential distribution, the task of finding an estimator for $(t + \delta)$ is more difficult. The values of the river flow before time t must be taken into account, contrary to the case of filtered Poisson process, for which the forecasts depend only on (t).

Moreover, because we want to clearly see the improvement obtained by using a more realistic distribution for the interarrival time, we will take the response function defined in (2). Indeed, as mentioned above, filtered Poisson processes having this response function are commonly used in hydrology, and the forecasting formulas are then easy to derive and implement. Therefore, it will be easier to judge the quality of the model that we propose.

In the next section, filtered renewal processes will be formally defined and the formulas needed to forecast the river flows will be developed. Then, in Section 3, the results will be applied to forecast the flows of the Delaware and Hudson Rivers. Finally, a few concluding remarks will be made in Section 4.

RIVER FLOWS MODELED AS A FILTERED RENEWAL PROCESS

Let (t) denote the flow of a certain river at time t. We assume that

$$X(t) = \sum_{n=1}^{N(t)} Y_n e^{-(t-\tau_n)/c} \quad (X(t) = 0 \text{ if } N(t) = 0),$$

$$(4)$$

in which $\{N(t), t \geq 0\}$ is a renewal process. The τ_n's are the arrival times of the events of the process $\{N(t), t \geq 0\}$, and the times $T_n = \tau_n - \tau_n - 1$ between the successive renewals are independent and identically distributed random variables. Finally, the (positive) random variables Y_1, Y_2,... are independent and identically distributed (and independent of $\{N(t), t \geq 0\}$).

Thus, the process is characterized by the sudden increases of the river flow caused by the events that occur at times $\tau_1, \tau_2, ...$ A renewal cycle is defined as the time T_n between two consecutive peaks. For the interarrival time, the following distributions (in addition to the exponential distribution) will be considered in Section 3: lognormal, gamma, Weibull, and Rayleigh.

Next, in order to be able to forecast the river flow at time $t+\delta$, we must first estimate the parameter c that appears in the response function. If there are no signals between t and $t+1$, then $(t + 1)$ is given by

$$X(t + 1) = e^{-1/c} X(t).$$

$$(5)$$

Hence, to estimate the parameter c, we can consider all the days where the flow decreased in the data set and calculate the ratio $(t+1)/X(t)$. The point estimate of c is then derived from the arithmetic mean $\overline{R}$ of this ratio:

$$\hat{c} = -\frac{1}{\ln \overline{R}}.$$

$$(6)$$

In the case of the random variables Y_n, which constitute a random sample of a parent random variable Y, their exact distribution is not needed. Indeed, only the expected value of Y will be used to forecast the value of $(t + \delta)$. Therefore, we simply have to compute the mean increase of the river flow during the calibration period of the model.

We will now derive formulas to estimate the river flow one and two days in advance. These formulas can be used with any distribution for the interarrival time T, which denotes the parent variable of the T_n's.

Forecasting the Flow at Time $t+1$

We will first try to forecast the river flow at time $t+1$, given all the information available up to t. To do so, we assume that there will be at most one event in the interval $(t, t + 1]$. At any rate, as mentioned above, in practice flow values are generally available on a daily basis. Therefore, it is not really possible to determine whether more than one event occurred in $(t, t + 1]$. When it does happen, these events can be considered as a single large event.

Apart from the case when the interarrival time T has an exponential distribution, we must take the history of the process into account to calculate the probability that there will indeed be an event in $(t, t + 1]$. Let

$$U_i = \tau_{N(t)+i} - \tau_{N(t)+i-1} \quad \text{for } i = 1, 2, \ldots.$$

$$(7)$$

That is, U_i is the time elapsed between the $(i-1)$st and the ith event after the most recent one that occurred before or at time t. The random variables T and U_1 are identically distributed. We must compute

$$p_k := P\left[U_1 \le k+1 \mid U_1 > k\right],$$

(8)

where k is the (integer) number of days since the most recent signal was observed. We have

$$p_k = \frac{\int_k^{k+1} f_{U_1}(u)\,du}{\int_k^{\infty} f_{U_1}(u)\,du}.$$

(9)

With the help of a mathematical software, we can obtain numerical values of this probability for any density function f_{U_1} and any value of k.

Next, if T is exponentially distributed, it is well known that, given that there is a signal (i.e., an event) between t and $t+1$, the conditional density of T is uniform on the interval $(t, t+1]$, so that the signal in question occurred on average at time $t+1/2$. In the general case, we must compute the conditional density function of U_1 in the interval $(k, k+1]$, given that $U_1 > k$.

If there is indeed a signal between t and $t+1$, the forecast of the river flow at time $t+1$ would be

$$Z_k := X(t)\,e^{-1/\hat{c}} + E[Y]\,\frac{1}{P[U_1 > k]}$$

$$\times \int_k^{k+1} e^{-(k+1-u)/\hat{c}} f_{U_1}(u)\,du.$$

(10)

Therefore, the forecast of the river flow at time $t+1$, given the entire history of the process up to t, is given by

$$\widehat{X}(t+1) = X(t)\,e^{-1/\hat{c}}\left(1 - p_k\right) + Z_k p_k$$

$$= X(t)\,e^{-1/\hat{c}} + p_k E[Y]\,\frac{1}{P[U_1 > k]}$$

$$\times e^{-(k+1)/\hat{c}} \int_k^{k+1} e^{u/\hat{c}} f_{U_1}(u)\,du.$$

(11)

Forecasting the Flow at Time $t+2$

Let $U_{1,} = U_1 \mid \{U_1 > k\}$. To forecast the river flow at time $t+2$, given all the information available up to time t, we need the distribution of the random variable $S := U_{1,k} + U_2$, where U_2 is defined in (7).

We now assume that the probability that there will be more than two signals in the interval $(t, t+2]$ is negligible (irrespective of the value of k). The density function of S_k is given by the convolution of the density functions of $U_{1,}$ and U_2:

$$
f_{S_k}(s+k) = \int_k^\infty f_{U_{1,k}}(u) f_{U_2}(s+k-u)\, du.
$$

(12)

Since U_2 is a nonnegative random variable, we can write that

$$
f_{S_k}(s+k) = \int_k^{s+k} f_{U_{1,k}}(u) f_{U_2}(s+k-u)\, du \quad \text{for } s > 0.
$$

(13)

In the case when T has an exponential distribution with parameter λ, we know that $(S_k - k)$ has a gamma distribution with parameters $\alpha=2$ and λ. In the other cases, we must generally evaluate the above integral numerically.

The value of $(t+2)$ is given by

$$
X(t+2) = \begin{cases}
e^{-2/c}X(t) & \text{if } U_{1,k} > k+2, \\
e^{-2/c}X(t) + Y_1^* e^{-(k+2-U_{1,k})/c} & \text{if } U_{1,k} \le k+2 \\
& \text{and } U_2 > 2 - (U_{1,k} - k), \\
e^{-2/c}X(t) + Y_1^* e^{-(k+2-U_{1,k})/c} & \text{if } U_{1,k} + U_2 \le k+2 \\
\quad + Y_2^* e^{-(2-U_2)/c} & (\text{and } U_3 > 2 - (U_{1,k} - k) - U_2),
\end{cases}
$$

(14)

where Y_i^* denotes the size of the ith signal in the interval $(t, t+2]$. As mentioned above, we assume that

$$
P\left[U_3 > 2 - (U_{1,k} - k) - U_2\right] \simeq 1.
$$

(15)

We must first compute

$$
p_{1,k} := P\left[U_{1,k} \in (k, k+2]\right].
$$

(16)

When U_1 has an exponential distribution with parameter λ,

$$
p_{1,k} = P\left[U_1 \in (0, 2]\right] = 1 - e^{-2\lambda}.
$$

(17)

In general, we have

$$P_{1,k} = \frac{\int_k^{k+2} f_{U_1}(u)\, du}{\int_k^{\infty} f_{U_1}(u)\, du}.$$

(18)

Next,

$$P_{2,k} := P\left[U_{1,k} + U_2 \in (k, k+2]\right] \equiv P\left[S_k \in (k, k+2]\right]$$

$$= \int_0^2 f_{S_k}(s+k)\, ds$$

$$= \int_0^2 \left[\int_k^{s+k} f_{U_{1,k}}(u)\, f_{U_2}(s+k-u)\, du\right] ds.$$

(19)

Alternatively, using the fact that $U1$, and $U2$ (≥ 0) are independent random variables, we may write that

$$P_{2,k} := P\left[U_{1,k} + U_2 \in (k, k+2]\right]$$

$$= \int_k^{k+2} P\left[U_{1,k} + U_2 \in (k, k+2] \mid U_{1,k} = u\right] f_{U_{1,k}}(u)\, du$$

$$= \int_k^{k+2} P\left[U_2 \in (k-u, k+2-u]\right] f_{U_{1,k}}(u)\, du$$

$$= \int_k^{k+2} \left[\int_0^{k+2-u} f_{U_2}(v)\, dv\right] f_{U_{1,k}}(u)\, du.$$

(20)

Finally, by independence,

$$P\left[\{U_{1,k} \leq k+2\} \cap \{U_2 > 2 - (U_{1,k} - k)\}\right]$$

$$= \int_k^{k+2} P\left[U_2 \in (k+2-u, \infty)\right] f_{U_{1,k}}(u)\, du$$

$$= \int_k^{k+2} \left[\int_{k+2-u}^{\infty} f_{U_2}(v)\, dv\right] f_{U_{1,k}}(u)\, du$$

$$= \int_k^{k+2} \left[1 - \int_0^{k+2-u} f_{U_2}(v)\, dv\right] f_{U_{1,k}}(u)\, du$$

$$= P_{1,k} - P_{2,k}.$$

(21)

The forecast of the river flow at time $t+2$, given the entire history of the process up to time t, is thus given by

$$\widehat{X}(t+2)$$

$$= X(t)\, e^{-2/\hat{c}}\,(1 - p_{1,k})$$

$$+ \left\{ X(t)\, e^{-2/\hat{c}} + E[Y] \int_k^{k+2} e^{-(k+2-u)/\hat{c}} f_{U_{1,k}}(u)\, du \right\}$$

$$\times (p_{1,k} - p_{2,k})$$

$$+ \left\{ X(t)\, e^{-2/\hat{c}} + E[Y] \int_k^{k+2} e^{-(k+2-u)/\hat{c}} f_{U_{1,k}}(u)\, du \right.$$

$$+ E[Y] \int_k^{k+2} \left[\int_0^{k+2-u} e^{-(2-v)/\hat{c}} f_{U_2}(v)\, dv \right]$$

$$\left. \times f_{U_{1,k}}(u)\, du \right\} p_{2,k}$$

$$= X(t)\, e^{-2/\hat{c}} + \left\{ E[Y] \int_k^{k+2} e^{-(k+2-u)/\hat{c}} f_{U_{1,k}}(u)\, du \right\} p_{1,k}$$

$$+ \left\{ E[Y] \int_k^{k+2} \left[\int_0^{k+2-u} e^{-(2-v)/\hat{c}} f_{U_2}(v)\, dv \right] \right.$$

$$\left. \times f_{U_{1,k}}(u)\, du \right\} p_{2,k}$$

$$= X(t)\, e^{-2/\hat{c}} + \left\{ E[Y]\, e^{-(k+2)/\hat{c}} \int_k^{k+2} e^{u/\hat{c}} f_{U_{1,k}}(u)\, du \right\} p_{1,k}$$

$$+ \left\{ E[Y] \int_k^{k+2} \left[e^{-2/\hat{c}} \int_0^{k+2-u} e^{v/\hat{c}} f_{U_2}(v)\, dv \right] \right.$$

$$\left. \times f_{U_{1,k}}(u)\, du \right\} p_{2,k}.$$

$$(22)$$

FORECASTING THE FLOWS OF THE DELAWARE AND HUDSON RIVERS

To assess the quality of the forecasts obtained with the renewal filtered process, we need to apply the formulas developed in the previous section to real-life data. As in [7], we chose the Delaware and Hudson Rivers, located in the United States, at the Montague (01438500) and the North Creek (01315500) gage stations, respectively. The flow values are available on the Internet at the following address: http://nwis.waterdata.usgs.gov. To calibrate the model, we used the data from October 2008 to September 2009.

Remarks. (i) When one is dealing with real-life data, things are not as simple as the mathematical models suggest. Looking at the actual hydrograph of the Delaware River, we observe a number of weak peaks that occur while the flow is decreasing and for which the increasing period lasts only one day before the flow resumes its decline. Similarly, there are sometimes small increases of the flow value occurring just after a minimum was observed that last only one day. In building the data set, there is always a subjective part. We decided to neglect these weak peaks, thus considering only the peaks that appear quite clearly in the hydrograph.

(ii) Because the river flow is observed on a daily basis, we must discretize the set of possible values taken by the interarrival time T. We computed the number k of days elapsed between consecutive peaks in the data set. In the case of the first observed peak, k is the number of days elapsed since the beginning of the calibration period. Then, k is obtained by subtracting the arrival times of the consecutive peaks. Notice that k is greater than or equal to 1.

We first consider the Delaware River. For this river, we found that the average value $\overline{T}$ of the interarrival times in the data set is equal to 6.8889 days. Moreover, the standard deviation of the observations is given by $s_T = 3.8977$ days. Based on these values, one may at once conclude that T is very unlikely to be exponentially distributed. Indeed, remember that the mean and the standard deviation of an exponential distribution with parameter λ are both equal to $1/\lambda$.

Next, the value of the point estimate of the (unitless) parameter c (see (6)) is $\hat{c} = 9.2592$, and the average value of the magnitude Y of the signals is given by $\overline{Y} = 984.6121$ cubic feet per second.

For the distribution of the random variable T, we tested the following models:

- An exponential distribution with parameter λ,
- A lognormal distribution with parameters μ and σ,

- A gamma distribution with shape parameter α and scale parameter λ,
- A weibull distribution with shape parameter λ and scale parameter κ,
- A rayleigh distribution with parameter σ.

Table 1 summarizes the results of the chi-square goodness-of-fit tests for each distribution and gives the respective parameter estimates. We find that, as expected, the exponential distribution is rejected, whereas the other distributions are all accepted, according to the P-values of the tests. The empirical cumulative distribution function as well as the distribution functions of the various models considered above is shown in Figure 2. We clearly see that the exponential distribution does not fit the data nearly as well as the other four models.

Table 1: Goodness-of-fit tests for the distribution of the random variable T (Delaware River).

Distribution	Exponential	Lognormal	Gamma	Weibull	Ray-leigh
Parameters	1/6.8889	1.7574, 0.6187	3.0544, 2.2554	7.7974, 1.8975	5.5817
P-value	0.0344	0.7887	0.7723	0.6927	0.7008

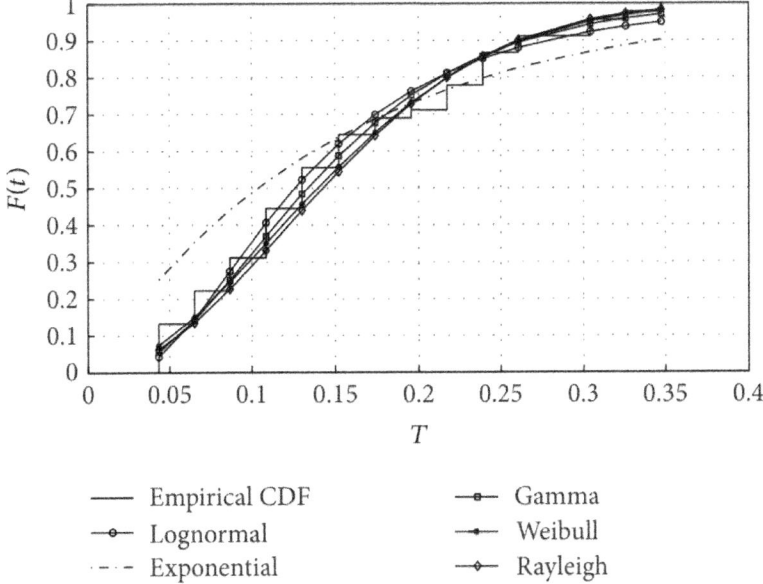

Figure 2: Empirical and fitted distribution functions for the random variable T (Delaware River).

Now, in the case of the filtered Poisson process (1) with the response function defined in (2), to estimate the river flow at time $t+1$ we compute the expected value of $X(t + 1)$ given all the past observations. Because of the memoryless property of the exponential distribution, we find that this conditional expectation only depends on the most recent value of the flow, that is, (t). We obtain (see [7]) that

$$E[X(t + 1) X] = (t) e^{-1/c} X(t) + E[X(1)], \tag{23}$$

where

$$E[X(1)] = \frac{\lambda c}{\mu} \left(1 - e^{-1/c}\right). \tag{24}$$

Similarly, to estimate the river flow at time $t+2$, we only need to calculate

$$E[X(t + 2) \mid X(t)] = E\left[\sum_{n:0 \leq \tau_n \leq t+2} Y_n e^{-(t+2-\tau_n)/c} \mid X(t) \right]$$

$$= e^{-2/c} E\left[\sum_{n:0 \leq \tau_n \leq t} Y_n e^{-(t-\tau_n)/c} \mid X(t) \right]$$

$$+ E\left[\sum_{n:t < \tau_n \leq t+2} Y_n e^{-(t+2-\tau_n)/c} \mid X(t) \right]$$

$$= e^{-2/c} X(t) + E[X(2)], \tag{25}$$

where

$$E[X(2)] = \frac{\lambda c}{\mu} \left(1 - e^{-2/c}\right). \tag{26}$$

Moreover, we have that

$$\lim_{t \to \infty} E[X(t)] = \frac{\lambda c}{\mu}, \tag{27}$$

$$\lim_{t \to \infty} \text{Var}[X(t)] = \frac{\lambda c}{\mu^2}, \tag{28}$$

$$\lim_{t \to \infty} \rho_{X(t),X(t+\delta)} = e^{-\delta/c}, \tag{29}$$

where $\rho_{X(t),X(t+\delta)}$ denotes the correlation coefficient of $X(t)$ and $X(t + \delta)$.

By making use of these three equations, we can estimate the parameter λ of the Poisson process, the parameter μ of the assumed exponential distribution of the Y_n's, and the parameter c in the response function. Actually, we do not need point estimates of λ and μ to forecast the flow. From (27), we deduce that we can simply estimate the ratio $\lambda c/\mu$ that appears in both (24) and (26) by the mean value of the observed data when the process is in steady state. Finally, to estimate the parameter c, we first compute the empirical correlation coefficient r_1 between the flow values at times t and $t+1$ and then we set $\hat{c} = -1/\ln(r_1)$ (see (29)). With the value of r_1 being close to 1, the previous formula is well defined.

Remark. If one is only interested in forecasting the river flow at time $t+\delta$, one may estimate c by computing the empirical value of $\rho_{X(t),X(t+\delta)}$ and making use of (29).

We are now ready to compare the forecasts derived from both the filtered Poisson process and the filtered renewal process with the various distributions considered for the interarrival time T. We used the formulas given above to forecast the flow of the Delaware River for the 89-day period from February to April 2010.

To assess the accuracy of the forecasts, we consider four criteria commonly used in hydrology: the mean absolute percentage error (Mape), the Nash criterion (Nash), the peak criterion (Pc), and the correlation coefficient (Corr) between the observed and forecasted values.

The Nash criterion is based on the mean square error and is used to assess the predictive power of hydrological models. It evaluates the quality of the forecasts from the differences between the expected and observed daily values. It is equal to 1 in case of a perfect fit. For its part, the peak criterion is used to measure the quality of the forecasts during the critical peak period. The closer to 0 it is, the better the forecasts are.

Table 2 displays the results obtained with the two competing models. Looking at these results, we notice that the Mape, Nash, and Corr criteria yield practically the same values for the filtered renewal and filtered Poisson processes. However, the filtered renewal process did much better than the filtered Poisson process when we consider the peak criterion. This criterion is especially important because accurate forecasts are really needed during the peak period.

Table 2: Criteria for comparing the forecasting models (Delaware River).

		Filtered renewal process				Filtered Poisson process
		Lognormal	Gamma	Weibull	Rayleigh	Exponential
$X(t+1)$	Mape	12.58%	12.49%	12.55%	12.56%	12.76%
	Nash	0.8288	0.8290	0.8287	0.8287	0.8356
	Pc	0.1674	0.1664	0.1663	0.1669	0.2522
	Corr	0.9188	0.9186	0.9180	0.9188	0.9187
$X(t+2)$	Mape	20.36%	20.33%	20.34%	20.34%	21.83%
	Nash	0.5447	0.5445	0.5445	0.5445	0.5573
	Pc	0.2930	0.2893	0.2869	0.2863	0.3344
	Corr	0.7735	0.7732	0.7731	0.7730	0.7734

We now turn to the Hudson River. The value of the point estimate of the constant c for this river is $\widehat{c} = 9.1172$ and the average magnitude of the signals is $\overline{Y} = 229.2533$, which is much smaller than in the case of the Delaware River.

Proceeding as above, we performed chi-square goodness-of-fit tests for the various distributions considered. The results are presented in Table 3 (see also Figure 3). This time, we see that both the exponential and Rayleigh distributions must be rejected, while the lognormal distribution clearly provides the best fit to the data.

Table 3: Goodness-of-fit tests for the distribution of the random variable T (Hudson River).

Distribution	Exponential	Lognormal	Gamma	Weibull	Rayleigh
Parameters	1/8.3488	1.8754, 0.6625	9.2134, 1.3399	2.1785, 3.8324	7.9978
P-value	0.0013	0.6983	0.3203	0.2532	0.000139

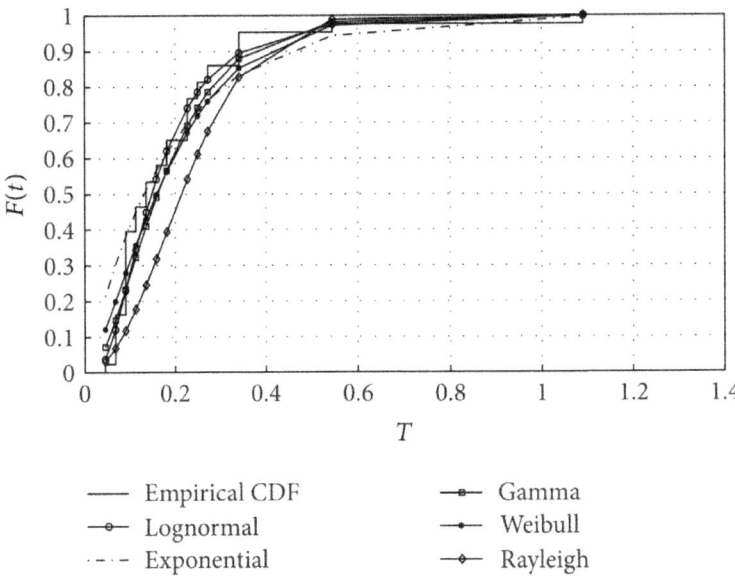

— Empirical CDF   —■— Gamma

—○— Lognormal   —•— Weibull

·—·— Exponential   —◆— Rayleigh

Figure 3: Empirical and fitted distribution functions for the random variable T (Hudson River).

The values of the four criteria used to compare the models are shown in Table 4. We discarded the Rayleigh distribution, because it yielded a very bad fit to the data. As in the case of the Delaware River, the values of the Nash and Corr criteria are quite similar, while the important peak criterion and the Mape criterion are much smaller for the filtered renewal process than for the filtered Poisson process.

Table 4: Criteria for comparing the forecasting models (Hudson River).

		Filtered renewal process			Filtered Poisson process
		Lognormal	Gamma	Weibull	Exponential
$X(t+1)$	Mape	12.33%	12.14%	12.33%	15.58%
	Nash	0.8240	0.8243	0.8240	0.8360
	Pc	0.2875	0.2895	0.2872	0.4818
	Corr	0.9146	0.9146	0.9146	0.9146

	Mape	20.84%	21.00%	21.15%	32.78%
	Nash	0.5841	0.5838	0.5835	0.6222
$X(t+2)$	Pc	0.4558	0.4533	0.4518	0.5575
	Corr	0.7908	0.7909	0.7910	0.7910

Finally, we present the observed and forecasted values of the flows of the Delaware and Hudson Rivers at times $t+1$ and $t+2$ in Figures 4, 5, 6, and 7, based on the filtered Poisson process (FPP) and the filtered renewal process (FRP) with a lognormal distribution for the interarrival time T. Notice that the filtered renewal process is generally better than the filtered Poisson process at forecasting high flow values, which is our main concern.

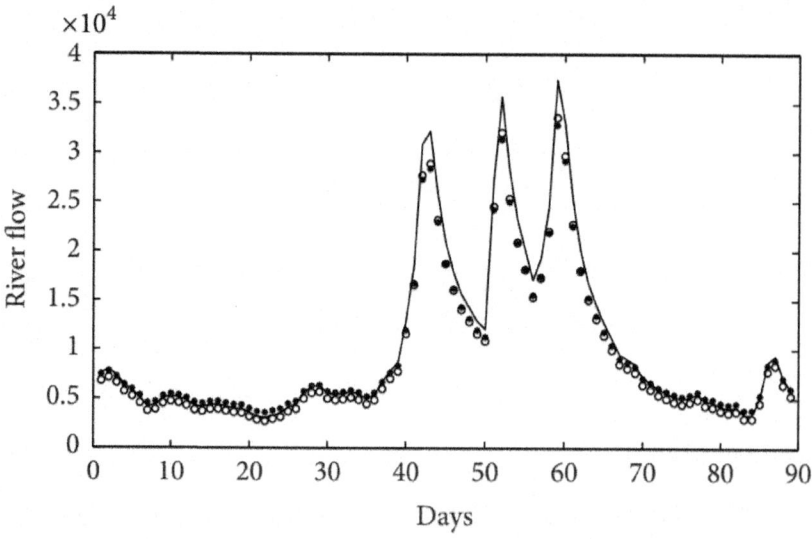

——— Observed value
  •   Predicted value (FPP)
  o   Predicted value (FRP)

Figure 4: Observed and forecasted values of the flow of the Delaware River at time $t+1$.

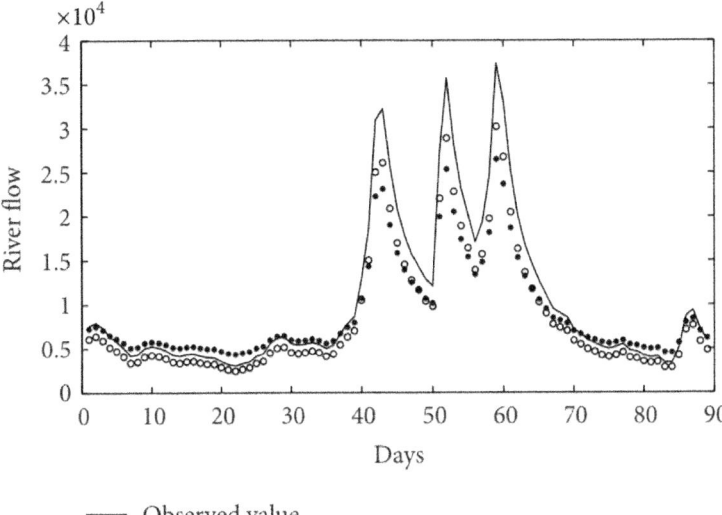

Figure 5: Observed and forecasted values of the flow of the Delaware River at time $t+2$.

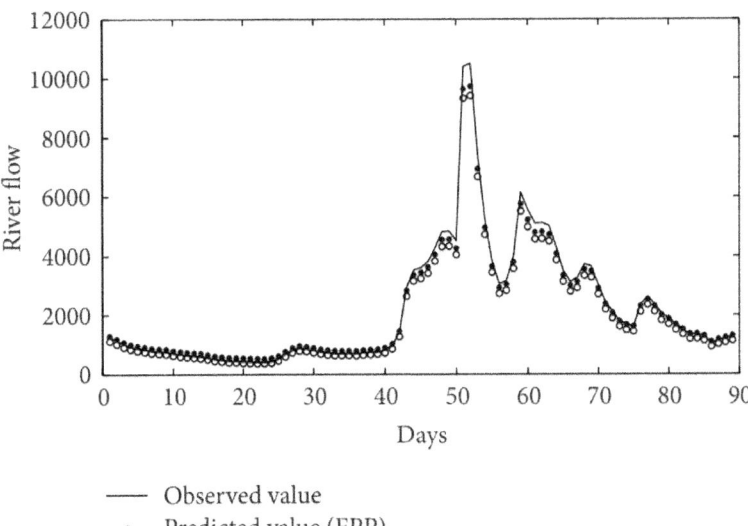

Figure 6: Observed and forecasted values of the flow of the Hudson River at time $t+1$.

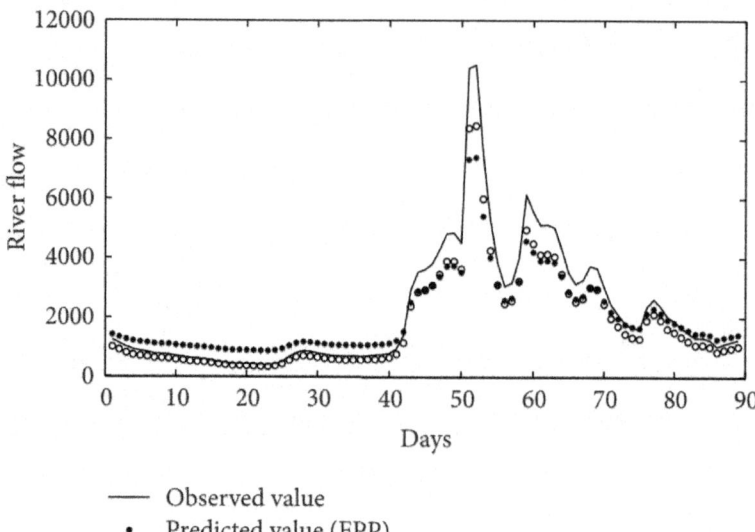

Figure 7: Observed and forecasted values of the flow of the Hudson River at time $t+2$.

CONCLUSION

In this paper, we were able to significantly improve the short-term hydrological forecasts produced by filtered Poisson processes by using a more realistic distribution for the interarrival time of the signals, thus working with filtered renewal processes. As we saw in the previous section, the filtered Poisson process, despite its lack of realism, is able to yield reasonable short-term forecasts. One possible explanation is the fact that the forecasting formulas derived from this model are very easy to implement. Also, contrary to the filtered renewal process, it does not require subjective decisions to estimate its various parameters. However, the filtered Poisson process was not able to produce forecasts as accurate as those derived from the filtered renewal process during the important peak period.

In addition to being used to forecast flow values, the models considered in this paper can also give us an estimate of the probability that the river flow will exceed a certain threshold in the next few days. This threshold may be a value that corresponds to a flow above which the risk of flooding is very high.

Finally, we could use the response function in (3) to further improve the forecasts. To do so, we would have to be able to estimate the parameter d in that response function and then obtain formulas that generalize the ones in (11) and (22), which is probably not an easy task.

ACKNOWLEDGMENT

The authors want to thank the anonymous reviewers for the constructive comments.

REFERENCES

1. G. Weiss, "Shot noise models for the generation of synthetic streamflow data," Water Resources Research, vol. 13, no. 1, pp. 101–108, 1977.

2. J. Kelman, "A stochastic model for daily streamflow," Journal of Hydrology, vol. 47, no. 3-4, pp. 235–249, 1980.

3. R. W. Koch, "A stochastic streamflow model based on physical principles," Water Resources Research, vol. 21, no. 4, pp. 545–553, 1985.

4. F. Konecny, "On the shot-noise streamflow model and its applications," Stochastic Hydrology and Hydraulics, vol. 6, no. 4, pp. 289–303, 1992.

5. Y.-J. Yin, Y. Li, and W. M. Bulleit, "Stochastic modeling of snow loads using a filtered Poisson process,"Journal of Cold Regions Engineering, vol. 25, no. 1, pp. 16–36, 2010.

6. H. Miyamoto, J. Morioka, K. Kanda et al., "A stochastic model for tree vegetation dynamics in river courses with interaction by discharge fluctuation impacts," Journal of Japan Society of Civil Engineers B1: Hydraulic Engineering, vol. 67, no. 4, pp. 1405–1410, 2012.

7. M. Lefebvre and J.-L. Guilbault, "Using filtered Poisson processes to model a river flow," Applied Mathematical Modelling, vol. 32, no. 12, pp. 2792–2805, 2008.

8. M. Lefebvre, "A filtered renewal process as a model for a river flow," Mathematical Problems in Engineering, vol. 2005, no. 1, pp. 49–59, 2005.

9. H. A. Loaiciga and R. B. Leipnik, "Stochastic renewal model of low-flow streamflow sequences,"Stochastic Hydrology and Hydraulics, vol. 10, no. 1, pp. 65–85, 1996.

10. A. Yagouti, I. Abi-Zeid, T. B. M. J. Ouarda, and B. Bobée, "Revue de processus ponctuels et synthèse de tests statistiques pour le choix d'un type de processus," Journal of Water Science, vol. 14, no. 3, pp. 323–361, 2001.

11. A. H. Marcus, "Traffic noise as a filtered Markov renewal process," Journal of Applied Probability, vol. 10, no. 2, pp. 377–386, 1973.

12. P. K. Andersen, "Filtered renewal processes with a two-sided impact function," Journal of Applied Probability, vol. 16, no. 4, pp. 813–821, 1979.

13. P. K. Andersen, "A note on filtered Markov renewal processes," Journal of Applied Probability, vol. 18, no. 3, pp. 752–756, 1981.

Chapter 2

WATERSHED-SCALE HYDROLOGICAL MODELING METHODS AND APPLICATIONS

Prem B. Parajuli[1] and Ying Ouyang[2]

[1] Department of Agricultural and Biological Engineering at Mississippi State University, Mississippi State, USA

[2] USDA-Forest Service Center for Bottomland Hardwoods Research, Mississippi State, USA

INTRODUCTION

Pollution of surface water with harmful chemicals and eutrophication of rivers and lakes with excess nutrients are serious environmental concerns. The U.S. Environmental Protection Agency (USEPA) estimated that 53% of the 27% assessed rivers and streams miles and 69% of the 45% assessed lakes, ponds, and reservoirs acreage in the nation are impaired (USEPA, 2010). In Mississippi, 57% of the 5% assessed rivers and streams miles are impaired (USEPA, 2010). These impairment estimates may increase when assessments of more water bodies are performed and water quality criteria are improved. The most common water pollution concerns in U.S. rivers and streams are sediment, nutrients (Phosphorus and Nitrogen) and pathogens. Hydrological processes can significantly impact on the transport of water quality pollutants.

Non-point source pollution from agricultural, forest, and urban lands can contribute to water quality degradation. Total Maximum Daily Loads (TMDLs) are developed by states to improve water quality. The TMDL requires identifying and quantifying pollutant contributions from each source to devise source-specific pollutant reduction strategies to meet applicable water quality standards. Commonly, water quality assessment at the watershed scale is accomplished using two techniques: (a) watershed monitoring and (b) watershed modeling. Watershed models provide a tool for linking pollutants to the receiving streams. Models provide quick and cost-effective assessment of water quality conditions, as they can simulate hydrologic processes, which are affected by several factors including climate change, soils, and agricultural management practices. However, methods used to develop a model for watersheds can significantly impact in the model outputs. Here several hydrological and water quality models are described. Case studies of two commonly used models with calibration and validation are provided with current and future climate change scenarios. This book chapter briefly reviews currently available hydrologic and water quality models, and presents model application case studies, to provide a foundation for further model development and watershed assessment studies.

REVIEW OF WATER QUALITY MODELS

Several useful hydrologic and water quality models are available today, each with diverse capabilities for watershed assessment. Many of these models are relevant to water quality goal assessment and implementation. Modeling of hydrology, sediment and nutrients has developed substantially, but advances have not always been consistent with the needs of the water quality goals program. Comprehensive education and training with model applications and case studies are needed for users to understand the potentials, limitations, and suitable applications of a model. Review of several hydrological models (e.g. SWAT, AnnAGNPS, HSPF, SPARROW, GLEAMS, WEPP, EFDC etc.) including models description and application within the U.S. or other countries are discussed.

SWAT Model

The SWAT model is developed and supported by the USDA/ARS. It is a physically based watershed-scale continuous time-scale model, which operates on a daily time step. The SWAT model can simulate runoff, sediment, nutrients, pesticide, and bacteria transport from agricultural watersheds (Arnold et al., 1998). The SWAT model delineates a watershed, and sub-divides that watershed in to sub-basins. In each sub-basin, the model creates several hydrologic response units (HRUs) based on specific land cover, soil, and topographic conditions. Model simulations that are performed at the HRU levels are summarized for the sub-basins. Water is routed from HRUs to associated reaches in the SWAT model. SWAT first deposits estimated pollutants within the stream channel system then transport them to the outlet of the watershed. The HRUs provide opportunity to include processes for possible spatial and temporal variations in model input parameters. The hydrologic module of the model quantifies a soil water balance at each time step during the simulation period based on daily precipitation inputs.

The SWAT model distinguishes the effects of weather, surface runoff, evapo-transpiration, crop growth, nutrient loading, water routing, and the long-term effects of varying agricultural management practices (Neitsch et al., 2005). In the hydrologic module of the model, the surface runoff is estimated separately for each sub-basin and routed to quantify the total surface runoff for the watershed. Runoff volume is commonly estimated from daily rainfall using modified SCS-CN method. The Modified Universal Soil Loss Equation (MUSLE) is used to predict sediment yield from the watershed. The SWAT model has been extensively applied for simulating stream flow, sediment yield, and nutrient modeling (Gosain et al., 2005; Vache et al., 2002; Varanou et al., 2002). The model needs several data inputs to represent watershed conditions which include: digital elevation model (DEM), land use land cover, soils, climate data. The SWAT model is an advancement of the Simulator for Water Resources in Rural Basins (SWRRB) and Routing Outputs to Outlet (ROTO) models. The SWAT model development was influenced by other models like CREAMS (Knisel, 1980), GLEAMS (Leonard et al., 1987), and EPIC (Williams et al., 1984; Neitsch et al., 2002).

The SWAT model has been recently applied to assess watershed conditions of the U.S. (Gassman et al., 2007; Parajuli et al., 2008; 2009; Parajuli 2010a; 2011; 2012; Chaubey et al., 2010) and internationally such as Ethiopia (Betrie et al., 2011); Kenya and northwest Tanzania (Dessu and Melesse, 2012); Bulgaria and Greece (Boskidis et al., 2012); and Australia (Githui et al., 2012).

ANNAGNPS

The AnnAGNPS model is a product of the USDA Agriculture Research Service (USDA-ARS) and the USDA Natural Resources Conservation Service (USDA-NRCS) to evaluate non-point source pollution from agriculture watersheds. Similar to the SWAT model, it is a physically based continuous and daily time step model used to simulate surface runoff, sediment, and nutrient yields (Cronshey and Theurer, 1998; Bingner and Theurer, 2003). The AnnAGNPS is considered an enhanced modification to the single event based Agricultural Non-Point Source (AGNPS) model (Young et al., 1989), as it retains many features of AGNPS (Yuan et al., 2001). Unlike AGNPS, the AnnAGNPS delineates watershed, sub-divides the watershed into small drainage areas with homogenous land use, soils, etc. The sub-areas are integrated and simulated surface runoff and pollutant loads through rivers and streams within the sub-areas and watershed, which is enhanced from the AGNPS.

The AnnAGNPS model utilizes and incorporates components or sub-components from several other models such as; Revised Universal Soil Loss Equation (RUSLE) model (Renard et al., 1997); Chemicals, Runoff, and Erosion from Agricultural Management Systems (CREAMS) model (Knisel, 1980); Groundwater Loading Effects on Agricultural Management Systems (GLEAMS) model (Leonard et al., 1987); and Erosion Productivity Impact Calculator (EPIC) Model (Sharpley and Williams, 1990). The AnnAGNPS model represents small watershed areas using a cell-based approach, with land and soil property characterization similar to SWAT model HRUs. Daily soil moisture contents are calculated using the Curve Number (CN) method, which help to quantify surface and subsurface flows. The AnnAGNPS model uses the RUSLE to estimate sediment yields.

Refereed AnnAGNPS model based evaluations have been applied predominantly to watersheds located in the U.S. (Yuan et al., 2011; 2002; Zuercher et al., 2011; Polyakov et al., 2007). However, the model also has been applied in other countries such as Mediterranian (Licciardello et al., 2011; 2007); Australia (Baginska et al., 2003), and China (Hua et al., 2012).

WEPP

The Water Erosion Prediction Project (WEPP) model is a product of USDA. The WEPP model is a process-based, distributed parameter, single storm and continuous based model used to predict surface flow and sediment yields from the hill slopes and small watersheds. WEPP allows simulation of the effects of crop, crop rotation, contour farming, and strip cropping. The WEPP model components includes weather generation, snow accumulation and melt, irrigation, infiltration, overland flow process, water balance, plant growth, residue management, soil disturbance by tillage, and erosion processes. The WEPP model considers sheet and rill erosion processes to predict erosion. The WEPP model incorporates modified water balance and percolation components from the SWRRB model (Williams and Nicks, 1985). The WEPP model utilizes and incorporates components or sub-components from several other models such as; EPIC (Williams et al., 1984); and CREAMS model (Knisel, 1980). The WEPP model has undergone continuous development since 1992 (1992-1995 with DOS version; 1997-2000 with window interface; 1999-2009 with Geo-WEPP ArcView/ArcGIS extensions; and 2001-present with web-browser interface; Flanagan et al., 2007; Foltz et al., 2011).

Refereed WEPP-model-based evaluations exist predominantly for agricultural fields or small watersheds located in the U.S. (Dun et al., 2010; Flanagan et al., 2007; Foltz et al., 2011). However, the WEPP has been applied in other countries such as China (Zhang et al., 2008).

GLEAMS

Groundwater Loading Effects of Agricultural Management Systems (GLEAMS) is a daily time-step, continuous, field-scale hydrological and pollutant transport mathematical model (Leonard et al., 1987). The GLEAMS model can simulate surface runoff, percolation, nutrient and pesticide leaching, erosion and sedimentation. The GLEAMS model requires several daily climate data including mean daily air temperature, daily rainfall, mean monthly maximum and minimum temperatures, wind speed, solar radiation and dew-point temperature data. The soil input parameters in the model can be obtained from the State Soil Geographic Database (STATSGO) or Soil Survey Geographic Database (SSURGO) soil data. Previous studies described the ability of GLEAMS model to predict nitrate transport process from the agricultural areas (Shirmohammadi et al., 1998; Bakhsh et al., 2000; Chinkuyu and Kanwar, 2001).

Refereed GLEAMS model applications have been published predominantly for field scale studies in the U.S. (Bakhsh et al., 2000; Chinkuyu et al., 2004). However, GLEAMS also has been applied in a few other countries, such as China (Zhang et al., 2008).

HSPF MODEL

The hydrological simulation program—FORTRAN (HSPF) is a product of U.S. Environmental Protection Agency (US-EPA), which is a comprehensive model used for modeling processes related to water quantity and quality in watersheds of various sizes and complexities (Bicknell et al. 2001). It simulates both the land area of watersheds and the water bodies. The HSPF model uses input data including hourly history of rainfall, temperature and solar radiation; land surface characteristics/land use conditions; and land management practices to predict parameters at watershed scales. The results of model simulations are based on a time history of the quantity and quality of runoff from an urban, forest or agricultural watershed, which include surface runoff, sediment load, nutrients and pesticide concentrations. The HSPF model can simulate three sediment types (sand, silt, and clay) in addition to organic chemicals and alternative products. A detailed description of HSPF model can be found inBicknell et al. (2001).

There have been hundreds of applications of HSPF around the world (Bicknell et al., 2001; Akter and Babel, 2012; Ouyang et al., 2012; Rolle et al., 2012). Examples include applications in a large watershed at the Chesapeake Bay, in a small watershed near Watkinsville, GA, with the experimental plots of a few hectares and in other areas such as Seattle, WA, Patuxent River, MD., and Truckee-Carson Basins, NV. Details are available at: (http://water.usgs. gov/cgi-bin/man_wrdapp?hspf).

SPARROW

The SPAtially-Referenced Regression On Watershed attributes (SPARROW) model is a watershed modeling tool for comparing water-quality data collected at a network of monitoring stations to characterize watersheds containing the stations (Smith et al., 1997; Schwarz et al., 2008). The SPARROW model has a nonlinear regression equation depicting the non-conservative transport of contaminants from the point and diffuse sources on land surfaces to streams and rivers. The SPARROW predicts contaminant flux, concentration, and yield in streams. It has been used to evaluate alternative hypotheses about important contaminant sources and watershed properties that control contaminant load and transport over large spatial scales. The SPARROW can be used to explain spatial patterns of stream water quality in relation to human activities and natural processes.

Numerous applications of SPARROW have been performed to assess water quality in watersheds in recent years. Brown (2011) investigated nutrient sources and transport in the Missouri River Basin with SPARROW. Saad et al. (2011) applied SPARROW to estimate nutrient load and to improve water quality monitoring design using a multi-agency dataset. Alam and Goodall (2012) examined the effects of hydrologic and nitrogen source changes on nitrogen yield in the contiguous United States with SPARROW.

EFDC

The Environmental Fluid Dynamics Code (EFDC) is a multifunctional surface water modeling system, which includes hydrodynamic, sediment-contaminant, and eutrophication components (Hamrick, 1996) and is available to the public through US-EPA website available at: http://www.epa.gov/ceampubl/swater/efdc/index.html. The EFDC can be used to simulate aquatic systems in multiple dimensions with the stretched or sigma vertical coordinates and the Cartesian (or curvilinear), and orthogonal horizontal coordinates to represent the physical characteristics of a water body. A dynamically-coupled transport process for turbulent kinetic energy, turbulent length scale, salinity and temperature are included in the EFDC model. The EFDC allows for drying and wetting in shallow water bodies by a mass conservation scheme.

Refereed EFDC-model-based evaluations exist predominately for stream ecosystems. Examples include a three-dimensional hydrodynamic model of the Chicago River, Illinois (Sinha et al., 2012); the effect of interacting downstream branches on saltwater intrusion in the Modaomen Estuary, China (Gong et al., 2012); and comparison of two hydrodynamic models of Weeks Bay, Alabama (Alarcon et al., 2012).

SWMM

The US-EPA's Storm Water Management Model (SWMM) was initially developed in 1971, and has been significantly upgraded (http://www.epa.gov/

nrmrl/wswrd/wq/models /swmm/index.htm). The SWMM model is a widely used model for planning, analysis and design related to storm water runoff, sewers, and other drainage systems in urban areas. SWMM can simulate single storm-events or provide continuous prediction of surface-runoff quantity and quality from urban areas. In addition to predicting surface-runoff quantity and quality, the model can also predict flow rate, flow depth, and water quality in each pipe and channel.

There have been numerous applications of SWMM in the literature recently. Blumensaat et al. (2012) investigated sewer transport with SWMM under minimum data requirements. Cantone and Schmidt (2011) applied SWMM to improve understanding of the hydrologic response of highly urbanized watershed catchments like the Illinois Urban areas. Talei and Chua (2012) estimated the influence of lag-time on storm event-based hydrologic impacts (e.g. rainfall, surface-runoff) using the SWMM model and a data-driven approach.

METHODS TO DEVELOP A MODEL

Appropriate methods are needed to develop a model, utilize different data sources (e.g. digital elevation, soil, land use, weather etc.), and develop methods to quantify pollutants source loads in the model. As examples, the methods development process is described here for two commonly used models (i.e., SWAT and HSPF).

SWAT Model

The SWAT model utilizes digital elevation model (DEM), soils, land cover, and weather data such as precipitation, temperature, wind speed, solar radiation, and relative humidity. SWAT delineates watershed boundary and topographic characteristics of the watershed using National Elevation Dataset called digital elevation model (DEM) data, which are available in the grid form with different resolutions (e.g. 30m x 30m grid; 10m x 10m grid) generally collected by U.S. Geological survey (USGS, 1999) or other sources. The 30m grid data are commonly used in the large scale watershed modeling work. However, small watershed or field scale modeling may benefit from using of 10m x 10m resolution DEM data. Model defines land use inputs in the model are described using distributed land cover data (USDA-NASS, 2010) or other land use data. The time-specific land-cover data (e.g. 1992, 2001 and 2006) for the U.S. and Puerto Rico can be downloaded from the National Land Cover Database (NLCD), a publicly available data source. The distributed land cover data with land use classifications can provide essential model input for the watershed assessment. Currently, land-use data layers are available in geographic information systems (GIS) format, which is applicable for the watershed modeling.

The SWAT model also requires distributed detail soils data, which is available from either State Soil Geographic (STATSGO) database or Soil Survey Geographic (SSURGO) databases (USDA, 2005). The SSURGO database is the most detailed data source currently available in the U.S. as it provides more soil polygons per unit area. The DEM, soils, and landuse geographic data layers should be all projected in one projection system (e.g. Universal Transverse Mercator-UTM 1983, zone 16).

Most of the watershed or field scale models (e.g. SWAT, WEPP) have embedded weather stations and climate generators. However, more field-specific climate inputs (e.g. rainfall; daily minimum, maximum and mean temperatures; solar radiation; relative humidity, and wind speed) can be allowed in the model for the watershed assessment. Weather data such as daily rainfall and ambient temperature can be downloaded from the National Climatic Data Center (NCDC, 2012). Other field-specific model input parameters such as irrigation (e.g. auto or manual irrigation), fertilizer application (application rates, fertilizer type), crop rotation (e.g. corn after soybean), tillage (e.g. conventional, reduced, no-tillage), planting and harvesting dates can be defined (Parajuli, 2010b).

HSPF Model

The major procedures in water quality modeling with HSPF are the construction of a conceptual model, mathematical description of the conceptual model, preparation of input data such as time series parameter values, calibration and validation of the model, and application of the model for field conditions. Time series input data can be supplied into the HSPF model by using a stand-alone program or the Watershed Data Management program (WDM) provided in BASINS (Better assessment science integrating point and nonpoint sources). BASINS is a multipurpose environmental analysis system model, which can be utilized by regional, state, and local agencies for conducting water quality based studies. The BASINS system incorporates an open source geographic information system (GIS) (i.e., MapWindow), the national watershed and meteorological data, and the state-of-the-art environmental models such as HSPF, Pollutant Loading Application (PLOAD), and SWAT into one convenient package (USEPA, 2010).

Normally, the development of a HSPF model starts with a watershed delineation process, which includes the setup of digital elevation model (DEM) data in the ArcInfo grid format, generation of stream networks in shape format, and designation of watershed inlets or outlets using the watershed delineation tool built in the BASINS. The HSPF also needs land use and soil data to determine the area and the hydrologic parameters of each land use pattern in the model, which can be done with the land use and soil classification tool in the BASINS. The HSPF is a lumped parameter model with a modular structure. The PERLAND modular represents the pervious land segments over which a considerable amount of water infiltrates into the ground. The IMPLND modular denotes the Impervious land segments over which infiltration is negligible such as paved urban surfaces. Processes involving water bodies like streams and lakes are represented with the RCHRES module. These modules have many components dealing with hydrological and water quality processes. Detailed information about the structure and functioning of these modules can be found in elsewhere (Donigian and Crawford 1976;Donigian et al. 1984; Bicknel et al. 1993; Chen et al. 1998).

MODEL APPLICATION

Two watersheds in Mississippi (Upper Pearl River and Yazoo River Basin) were selected for modeling case studies using two hydrologic and water quality models (SWAT and HSPF). Models were calibrated and validated using USGS observed streamflow data for the current conditions and models were applied to predict future climate change scenarios impact on hydrology. Case studies demonstrated how future climate change scenarios impact streamflow from the watersheds.

SWAT Model

The main objective of this case study was to quantify the potential impact of future climate change scenarios on hydrologic characteristics such as monthly average streamflow with in the Upper Pearl River Watershed (UPRW) using the SWAT model. The specific objectives were to: (1) develop a site-specific SWAT model for the UPRW based on watershed characteristics, climatic, and hydrological conditions; (2) calibrate and validate model using USGS observed stream flow data; and (3) develop future climate change scenarios and quantify their impacts on stream flows.

Study Area and Model Development

The SWAT model was developed and applied in the UPRW (7,588 km²), which is located in Mississippi (Fig. 1). The UPRW covers ten counties (Choctaw, Attala, Winston, Leake, Nesobha, Kemper, Madison, Rankin, Scott and Newton) in Mississippi with predominant land uses of woodland (72%), grassland (20%), urban land (6%) and others (2%).

To develop the SWAT model, this case study utilized national elevation data, which is also called DEM data of 30m x 30m grids to delineate watershed boundary. The STATSGO was used to create distributed soil data input in the model. The land cover data was created using the cropland data layer in the model. The climate data (e.g. daily precipitation, temperature) were used from several weather stations within or near the watershed as maintained by the National Climatic Data Center. The SWAT model allows several potential evapotranspiration estimation method alternatives (e.g. Penman-Monteith, Hargreaves, Priestley-Taylor). This case study utilized the Penman-Monteith method to estimate PET, which requires daily rainfall, maximum and minimum temperatures, relative-humidity, solar radiation, and wind speed data. The additional data needed to simulate the SWAT model using Penman-Monteith PET method were generated by the SWAT model.

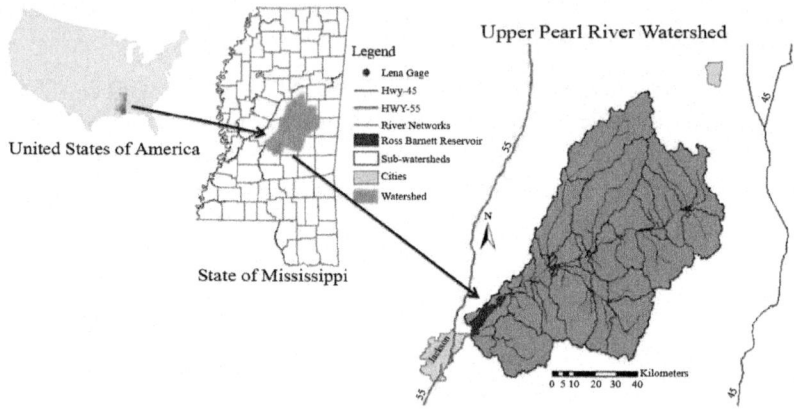

Figure 1. Location map of the watershed showing sub-watersheds and others

Model Calibration and Validation

The SWAT model predicted monthly streamflow values were compared separately for model calibration and validation periods using three common parameters (coefficient of determination – R^2; Nash–Sutcliffe efficiency index – E; and root mean square error - RMSE). The monthly model performances were ranked excellent for R^2 or E values > 0.90, very good for values between 0.75–0.89, good for values between 0.50–0.74, fair for values between 0.25–0.49, poor for values between 0–0.24, and unsatisfactory for values < 0 (Moriasi et al., 2007; Parajuli et al., 2008, 2009). The RMSE performance has no suggested values to rank, however the smaller the RMSE the better the performance of the model (Moriasi et al., 2007), and a value of zero for RMSE represents perfect simulation of the measured data.

The SWAT model was calibrated (from January 1998 to December 2003) and validated (from January 2004 to December 2009) using field observed monthly streamflow data from the Lena USGS gage station (USGS 02483500) within the UPRW. Model calibration and validation parameters were adopted from previous study (Parajuli, 2010a). Model simulated results showed good to very good performances for the monthly streamflow prediction both during model calibration (R^2 = 0.75, E = 0.70) and validation (R^2 = 0.73, E = 0.51) periods (Fig. 2). The SWAT model predicted monthly streamflow (m^3 s^{-1}) estimated very similar RMSE values (<2% difference) during model calibration (RMSE = 51.7 m^3 s^{-1}) and validation (RMSE = 50.7 m^3 s^{-1}) periods. This case study results were in close agreement with several previous studies that used the SWAT model (Gassman et al., 2007; Moriasi et al., 2007; Parajuli et al., 2009; Parajuli 2010a; Nejadhashemi et al., 2011; Sheshukov et al., 2011).

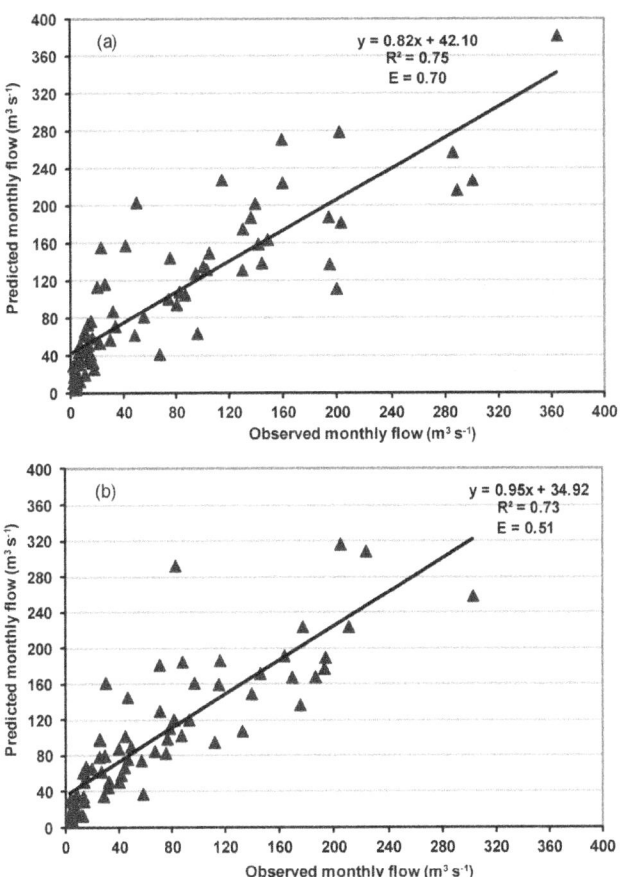

Figure 2. Monthly observed vs. predicted streamflow during (a) calibration and (b) validation periods

Future Climate Scenarios

The calibrated and validated SWAT model for the UPRW was simulated for an additional 30 years (January 2010 to December 2040) to provide fourteen future climate change scenarios (Table 1). The average streamflow value from the calibrated and validated model was considered as baseline scenario. The future climate change scenarios represented percentage change in the precipitation, temperature and CO_2 concentration values as described in the Table 1. The CO_2 values were adjusted from a baseline value of 330 ppmv (part per million by volume), which is a default value provided in the SWAT model. Two other CO_2 values (495 and 660) were tested in the model considering 50% and 100% increase from the model default value. Percentage changes in the precipitation were simulated for ±20% from the baseline value. Similarly, the model temperature factor was adjusted using +1 and +2 degrees in Celsius from the baseline. The fourteen future climate change scenarios were developed using interaction of three CO_2, three precipitation, and three temperature adjustment values.

The SWAT model results for fourteen scenarios (from Sc1 to Sc14 for Lena gage station) predicted an average maximum monthly stream flow decrease of 57% and average maximum monthly flow increase of 74% from the base simulation (Figure 3). Precipitation increase always had the greatest impact on monthly streamflow from the watershed. A twenty percent increase in precipitation resulted into the greatest impact in the future streamflow prediction. However, increases in CO_2 and temperature accelerated the magnitude of streamflow process.

Scenario 13 with the highest increase in the precipitation (+20%), CO_2 (660 ppmv), and temperature (+2 degree Celsius) had about 74% greater impact on streamflow prediction than the baseline condition (Fig. 3). Other scenarios that had high impact on streamflow prediction were Sc1, Sc4, Sc7, and Sc10. The increase in the temperature had medium impact on streamflow process as shown by Sc3, Sc6, Sc9, and Sc12. However, Sc12 had the greatest impact among medium scenarios as it predicted about 10% greater cumulative monthly streamflow than the baseline condition. Scenarios Sc2, Sc5, Sc8, Sc11, and Sc14 had lower cumulative monthly streamflow than the baseline condition, as they all had decreased precipitation (-20%). However, Sc14 had the greatest effect on stream flow among all low condition scenarios, due to the highest temperature (+2 degree Celsius) and CO_2 values (660 ppmv).

Table 1. Simulated climate change parameters scenarios and effect

CO$_2$ (ppmv)	Precip. (%)	Temp. (adj. °C)	Scenarios	Effect
330	0	0	Base	No
330	+20	0	Sc1	High
330	-20	0	Sc2	Low
330	0	+1	Sc3	Medium
330	+20	+1	Sc4	High
330	-20	+1	Sc5	Low
330	0	+2	Sc6	Medium
330	+20	+2	Sc7	High
330	-20	+2	Sc8	Low
495	0	+2	Sc9	Medium
495	+20	+2	Sc10	High
495	-20	+2	Sc11	Low
660	0	+2	Sc12	Medium
660	+20	+2	Sc13	High
660	-20	+2	Sc14	Low

CO$_2$ = carbon dioxide, Precip. = precipitation, Temp. = temperature, Sc = scenario

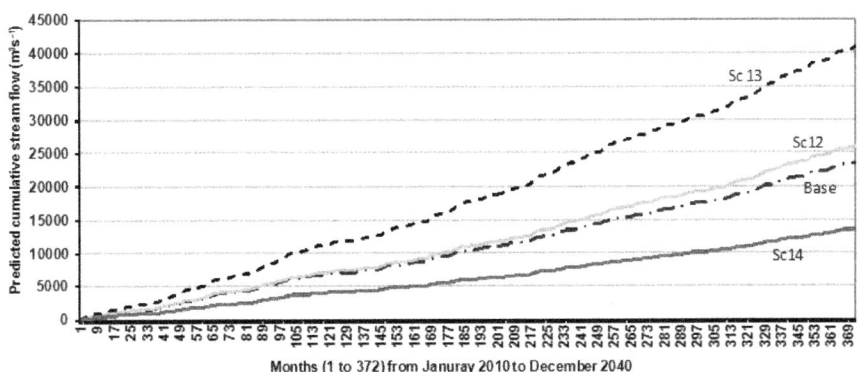

Figure 3. Model predicted cumulative monthly streamflow during thirty years period (2010-2040) showing greater than base condition and lower than base condition scenarios.

HSPF Model

The goal of this case study was to estimate the potential impact of future climate change upon hydrologic characteristics such as river discharge, surface evaporation, and water outflow in the YRB (Yazoo River Basin) using the HSPF model. The specific objectives were to: (1) develop a site-specific model for the YRB based on watershed, meteorological, and hydrological conditions;

(2) calibrate the resulting model using existing field data and/or computational data; and (3) create simulation scenarios to project the potential impact of future climate changes upon hydrologic characteristics in the YRB.

Study Area and Model Development

The YRB is the largest river basin in Mississippi, USA and has a total drainage area of 34,600 km^2(Fig. 4). This basin is separated into two distinct topographic regions, one is the Bluff Hills (about 16600 km^2) and the other is the Mississippi Alluvial Delta (Guedon and Thomas, 2004; MDEQ, 2008;Shields et al., 2008). The Bluff Hills region is a hilly and upland area where streams originate from lush oak and hickory forests and pastures dominate the rural landscape. The Delta Region, on the other hand, is a flat and lowland area characterized by slow streamflow and an extensive system of oxbow lakes.

Data collection for the YRB (HUC 8030208) includes watershed descriptions, meteorological, and hydrologic data. Several agencies are active in the data collection efforts. Most of the data used in this study such as land use, soil type, topography, precipitation, and discharge are from National Hydrography Dataset, U.S. Geologic Survey National Water Information System, and 2001 National Land Cover Data.

Four future climate change scenario data, namely the HADCM3B2, CSIROMK35A1B, CSIROMK2A1B, and MIROC32A1B, were used in this case study. HADCM3, CSIROMK35, CSIROMK2, and MIROC32 are names of climate general circulation models (GCM). The B2 and A1B at the end of the names of the climate change scenarios are the Intergovernmental Panel on Climate Change (IPCC) emission scenarios under which the GCMs were run to produce the individual climate projection. The HADCM3B2 scenario data was obtained from the Hadley Centre for Climate Prediction and Research, United Kingdom. The CSIROMK35A1B and CSIROMK2A1B scenarios data were obtained from the Australian Commonwealth Scientific and Research Organization Atmospheric Research, and the MIROC32A1B scenario data was obtained from the Center for Climate System Research, University of Tokyo National Institute for Environmental Studies and Frontier Research Center for Global Change. More detail information about these climate scenarios are available at: http://www-pcmdi.llnl.gov/ipcc/model_documentation/ ipcc_model_documentation.php. These four scenarios data involve monthly air temperature and precipitation for a period from 2000 to 2050, which were generated by GCMs and the Center for Climate System Research National Institute for Environmental Studies and Frontier Research Center for Global Change (University of Tokyo). These data were scaled to the 8-digit HUC watersheds for different regions. For the YRB watershed, the 8-digit HUC was 08030208. A descriptive statistics for these four scenarios data showed the amount of precipitation from high to low order as: CSIROMK35A1B > HADCM3B2 > CSIROMK2A1B > MIROC32A1B, whereas the magnitude of air temperature from high to low order as: MIROC32A1B > CSIROMK35A1B > CSIROMK32A1B > HADCM3B2.

The HSPF model for this case study was developed using the PERLND, IMPLND, and RCHRES modules that are available in HSPF. The PWATER section of the PERLND module is a major component that simulates the water budget, including surface flow, inter-flow and groundwater behavior. The HYDR section of the RCHRES module simulates the hydraulic behavior of the stream.

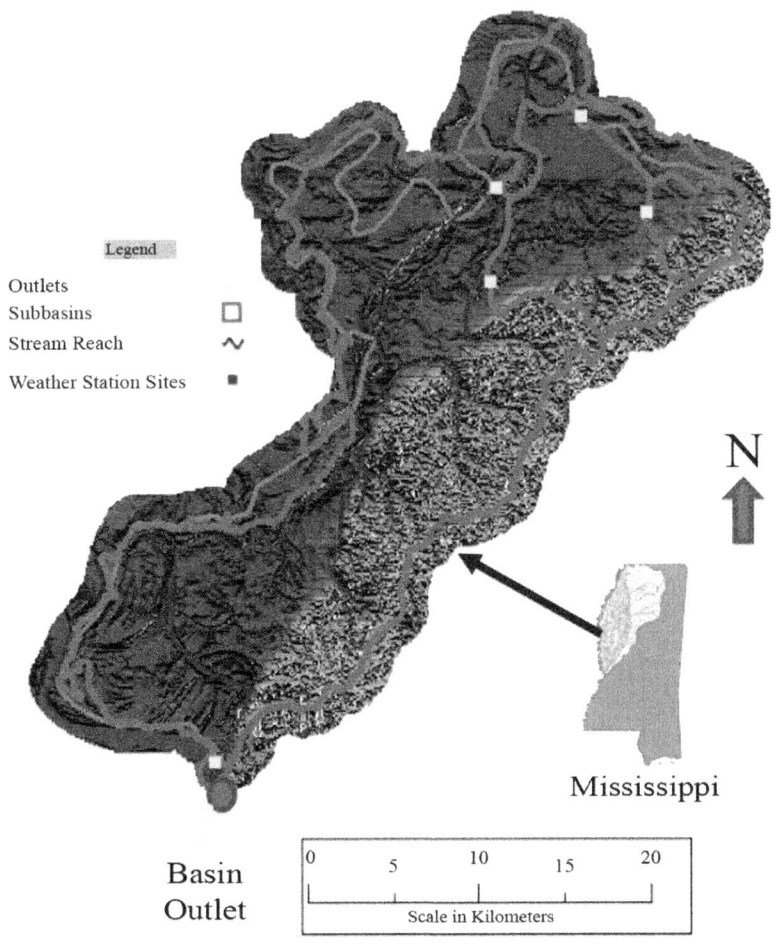

Figure 4. Location of modeled area in the Yazoo River Basin, Mississippi.

Model Calibration and Validation

Model calibration involves adjusting input parameters within a reasonable range to obtain a best fitness between field observations and model predictions. Model validation is a process of validating the calibrated model by comparing the field observations against the model predictions without changing any input parameter values. Table 2 shows a comparison of the observed and predicted annual water outflow volume. The annual differences in errors between the observed and predicted water outflow volumes were about 6% and were, therefore, acceptable (Bicknell et al., 2001). With prediction = 0.97*observation and R2 = 0.98 and E = 0.96, we determined that an excellent agreement was obtained between the field observations and model predictions during the model calibration process.

Comparison of annual water outflow between the observations and predictions for a time period from January 1, 2005 to December 31, 2010 during the model validation process was given in Table 2. The regression equation predictions = 0.97*observation and R^2 = 0.99 and E = 0.97 verified the excellent agreement between the model predictions and the field observations during the model validation process.

Table 2. Comparison of the simulated and observed annual water outflow volumes during model calibration and validation.

Year	Simulated Outflow (m^3)	Observed Outflow (m^3)	Percent Different
Model Calibration			
2000	1.77E+09	1.75E+09	0.88
2001	2.05E+09	1.92E+09	6.34
2002	1.93E+09	1.93E+09	-0.37
2003	1.15E+09	1.16E+09	-0.34
2004	2.00E+09	1.90E+09	5.58
Total	**8.90E+09**	**8.66E+09**	**2.68**
Model Validation			
2005	1.32E+09	1.30E+09	1.64
2006	1.33E+09	1.35E+09	-2.10
2007	1.20E+09	1.19E+09	1.13
2008	9.71E+08	9.47E+08	2.54
2009	1.96E+09	1.82E+09	7.40
Total	**6.78E+09**	**6.62E+09**	**2.50**

Past and Future Climate Change

Comparison of mean annual water yields between the past 10 years (2001-2011) and the future 40 years (2011-2050) for the four climate projections indicates that water yields will continue to decline (Table 3). The percent change in mean annual water yield varied from 29.47% for the CSIROMK35A1B projection to 18.51% for the MIROC32A1B projection, with four climate projections indicating continuing declines out to 2050. The same decline trends were observed for maximum annual water yields (Table 3). The declines in mean and maximum annual water yields occurred primarily due to the projected precipitation decrease. Mixed results were found for the mean annual evaporative loss (Table 3). The CSIROMK2B2 projection indicated a long-term increase while the other three projections indicated a long-term decrease in evaporative losses. Further work is thus necessary to better determine how evaporative losses will respond in the future.

Changes of monthly minimum, mean, and maximum in water discharges and yields for the four climate projections during the 40-year simulation period (2011-2050) are given in Figs. 5 and 6. The monthly minimum, mean, and maximum water discharges and yields varied among the four climate projections and changed from year to year within each projection. In general, the MIROC32A1B projection had highest monthly minimum, mean, and maximum water discharges and yields in most of the years during the 40-year simulation, which occurred because the MIROC32A1B projection had highest annual precipitation during the same simulation period (Table 3).

Table 3. Comparison of the sum and mean values for precipitation, evaporative loss, and water yield between the past and future 10 years.

Scenario	Precipitation (cm)			Evaporative Loss (m^3)			Water Yield (m^3)		
	Past 10 Years (2001 to 2010)	Future 40 Years (2011 to 2050)	% Change	Past 10 Years (2001 to 2010)	Future 10 Years (2011 to 2050)	% Change	Past 10 Years (2001 to 2050)	Future 10 Years (2011 to 2050)	% Change
Annual Mean									
HADCM3B2	0.017	0.015	-10.23	92.80	84.40	-9.95	40000.00	32282.00	-23.91
MIROC32A1B	0.016	0.015	-10.01	99.90	86.89	-14.97	37000.00	31222.00	-18.51
CSIROMK35A1B	0.019	0.017	-12.69	125.00	100.96	-23.81	53300.00	41169.00	-29.47
CSIROMK2B2	0.016	0.016	0.10	96.20	104.62	8.05	34800.00	32787.00	-6.14
Annual Maximum									
HADCM3B2	1.052	0.754	-39.45	484.00	330.25	-46.56	2160000.00	129412.00	-1569.09
MIROC32A1B	1.161	0.842	-37.79	618.00	372.10	-66.08	2150000.00	148377.00	-1349.01
CSIROMK35A1B	1.346	1.088	-23.73	580.00	438.07	-32.40	2160000.00	205767.00	-949.73
CSIROMK2B2	0.991	0.749	-32.33	488.00	371.37	-31.41	2160000.00	130590.00	-1554.03

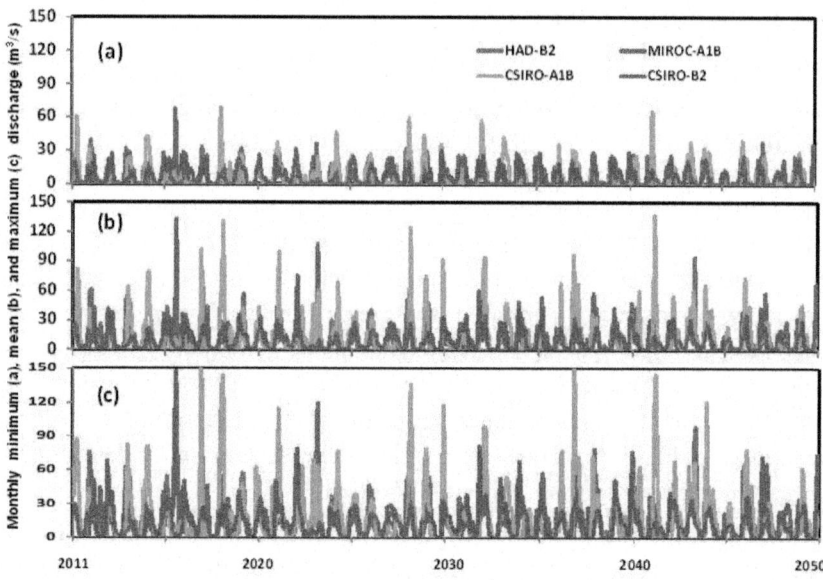

Figure 5. Simulated monthly minimum (a), mean (b), and maximum (c) discharge for the four simulation scenarios.

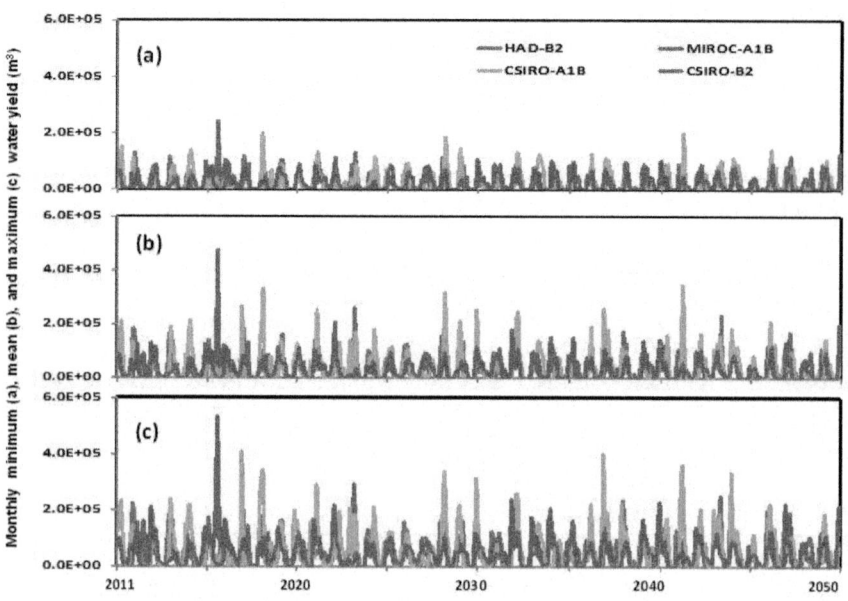

Figure 6. Simulated monthly minimum (a), mean (b), and maximum (c) water outflow volume for the four simulation scenarios.

CONCLUSIONS

Two models (SWAT and HSPF) commonly used in hydrological and water quality studies were applied here in two large scale watersheds (UPRW and YRB) in the state of Mississippi. Models were calibrated and validated using USGS observed streamflow data. The long-term hydrological impacts due to future climate change scenarios were assessed using the SWAT and HSPF models.

For one case study, simulated mean monthly streamflow results for the calibrated and validated SWAT model provided good to very good fits (R^2 and E values from 0.75 to 0.51) to USGS monthly observed streamflow data. Fourteen future climate change scenarios were developed using interaction of precipitation, CO_2, and temperature adjustment values in the SWAT model. The scenario with the highest increase in the precipitation (+20%), CO_2 (660 ppmv), and temperature (+2 degree Celsius) had about the greatest (> 74%) impact on streamflow simulation when compare with the baseline condition. Interaction of temperature adjustment and CO_2 factors had a medium and low impact respectively during thirty year's model simulation period in this study.

Another case study examined the impact of climate change on future water discharge, evaporation, and yield in the YRB using the BASINS-HSPF model. The model was calibrated using observed data from a five-year (2001 to 2004), and validated using observed data from another five-year (2005 to 2010). Excellent agreements were obtained between the model predictions and the field observations for model calibration and validation.

Four future climate scenarios (or projections) - CSIROMK35A1B, HADCM3B2, CSIROMK2B2, and MIROC32A1B were used to investigate water discharge, evaporative loss, and water outflow responses to predicted precipitation and air temperature changes over a 50-year period from 2001 to 2050. Comparison of simulation results between the past 10 years (2001-2010) and the future 40 years (2011-2050) shows that the mean and maximum annual water yields declined due to the projected precipitation decrease. In general, the MIROC32A1B projection had the highest monthly minimum, mean, and maximum water discharges and yields in most of the years during the 40-year simulation period (2011-2050). This projection had the highest projected annual precipitation. Results suggest that the projected precipitation had profound impacts upon water discharge and yield in the YRB.

Spatial data used in the models may have potential sources of errors. For example, the DEM data are used to delineate watershed boundary are available in different resolutions. Similarly, use of land use, soils and weather data may have some spatial errors, which can influence the hydrologic and climate change impact. However, these results will only have relative influence in model simulated results. This book chapter provided review of several watershed and water quality models and two case studies to evaluate future climate change impact on hydrology using two models.

ACKNOWLEDGEMENT

The part of the case study research presented in this book chapter is based on work supported by the Special Research Initiatives (SRI) and Mississippi Agricultural and Forestry Experiment Station (MAFES) at Mississippi State University.

REFERENCES

1.   Akter A., and Babel M. S. 2012. Hydrological modeling of the Mun River basin in Thailand. Journal of Hydrology, 452-453: 232-246.

2.   Alam M. J., and Goodall J. L. 2012. Toward disentangling the effect of hydrologic and nitrogen source changes from 1992 to 2001 on incremental nitrogen yield in the contiguous United States. Water Resources Research, 48 (4), W04506, doi:10.1029/2011WR010967 (in press).

3.   Alarcon V. J., McAnally W. H.,and Pathak S. 2012. Comparison of two hydrodynamic models of Weeks Bay, Alabama. Computational science and its applications – ICCSA, Lecture notes in Computer Science, 7334: 589-598, doi: 10.1007/978-3-642-31075-1_44.

4.   Arnold J. G., Srinivasan R., Muttiah R. S., and Williams J. R. 1998. Large area hydrologic modeling and assessment, Part I: model development. Journal of American Water Resources Association, 34(1): 73-89.

5.   Baginska B., Milne-Home W., and Cornish P. S. 2003. Modelling nutrient transport in Currency Creek, NSW with AnnAGNPS and PEST. Environmental Modelling & Software, 18: 801–808.

6.   Bakhsh A., R. S. Kanwar D. B. Jaynes T. S. Colvin and L. R. Ahuja. 2000. Prediction of NO3−N losses with subsurface drainage water from manured and UAN−fertilized plots using GLEAMS. Trans. ASAE, 43 (1): 69−77.

7.   Betrie G. D., Y. A. Mohamed A. van Griensven and R. Srinivasan. 2011. Sediment management modelling in the Blue Nile Basin using SWAT model. Hydrology and Earth System Sciences, 15: 807–818.

8.   Bicknell B. R., Imhoff J. C., Kittle J. L., Donigian A. S., and Johanson R. C. 1993. Hydrological Simulation Program – FORTRAN (HSPF): Users Manual for Release 10. EPA-600/R-93/174, U.S. EPA, Athens, GA, 30605

9. Bicknell, B. R., Imhoff, J. C., Kittle, Jr., J. L., Jobes, T. H., Donigian, Jr., A. S., 2001. Hydrological Simulation Program – Fortran, HSPF, Version 12, User's Manual. National Exposure Research Laboratory, Office of Research and Development, U.S. Environmental Protection Agency, Athens, GA.

10. Bingner R. L., and Theurer F. D. 2003. AnnAGNPS technical processes documentation, Version 3.2. USDA-ARS, National Sedimentation Laboratory: Oxford, MS.

11. Blumensaat F., Wolfram M., and Krebs P. 2012. Sewer model development under minimum data requirements. Environmental Earth Sciences, 65:1427-1437.

12. Boskidis, I., G. D. Gikas, G. K. Sylalos and V. A. Tsihrintzis. 2012. Water Resources Management, 26(10): 3023-3051.

13. Brown, J. B. 2011. Application of the SPARROW watershed model to describe nutrient sources and transport in the Missouri River Basin: U.S. Geological Survey Fact Sheet, 3104, 4 p.

14. Cantone, J., and Schmidt, A 2011. Improved understanding and prediction of the hydrologic response of highly urbanized catchments through development of the Illinois Urban Hydrologic Model. Water Resources Research, 47: W08538.

15. Chaubey, I., L. Chiang, M. W. Gitau, and S. Mohamed. 2010. Journal of soil and water conservation, 65 (6): 424-437.

16. Chen Y. D., Carsel R. F., Mccutcheon S. C., and Nutter W. L. 1998. Stream temperature simulation of forested riparian areas: I. Watershed model development. ASCE - Journal of Environ Engineering, 124:304–315

17. Chinkuyu, A. J., T. Meixner, T. Gish, and C. Daughtry. 2004. The importance of seepage zones in predicting soil moisture content and surface runoff using GLEAMS and RZWQM. Trans of the ASAE, 47(2): 427−438.

18. Chinkuyu, A. J., and R. S. Kanwar. 2001. Predicting soil nitrate−nitrogen losses from incorporated poultry manure using GLEAMS model. Trans. ASAE, 44 (6): 1643−1650.

19. Cronshey R. G., and Theurer F. G. 1998. AnnAGNPS-non point pollutant loading model. In Proceedings First Federal Interagency Hydrologic Modelling Conference. Las Vegas, NV.

20. Dessu, S. B. and A. M. Melesse. 2012. Modelling the rainfall-runoff process of the Mara River basin using the Soil and Water Assessment Tool. Hydrological Processes, DOI: 10.1002/hyp.9205.

21. Donigian A. S. Jr., and Crawford N. H. 1976. Modeling Pesticides and Nutrients on Agricultural Lands. Environmental Research Laboratory, Athens, GA. EPA 600/2-7-76-043, 317 p

22. Donogian A. S. Jr., Imhoff J. C., Bicknell B. R., Kittle J. I. 1984. Application guide for hydrological simulation program-FORTRAN (HSPF), EPA, Athens, GA. EPA-600/3-84-065.

23. Dun S., J. Q. Wu, D. K. McCool, J. R. Frankenberger, and D. C. Flanagan. 2010. Improving frost-simulation sub-routines of the Water Erosion Prediction Project (WEPP) model. Trans. of the ASABE, 53(5): 1399-1411.

24. Flanagan D. C., J. E. Gilley, and T. G. Franti. 2007. Water Erosion Prediction Project (WEPP): development history, model capabilities, and future enhancements. Trans. of the ASABE, 50(5): 1603-1612.

25. Foltz R. B., W. J. Elliot, and N. S. Wagenbrenner. 2011. Soil erosion model predictions using parent material/soil texture-based parameters compared to using site-specific parameters. Trans. of the ASABE, 54(4): 1347-1356.

26. Gassman P. W., Reyes M. R., Green C. H., and Arnold, J. G. 2007. The Soil and Water Assessment Tool: historical development, applications, and future research directions. Transactions of the ASABE, 50(4): 1211-1250.

27. Githui F., B. Selle and T. Thayalakumaran. 2012. Recharge estimation using remotely sensed evapotranspiration in an irrigated catchment in southeast Australia. Hydrological Processes, 26: 1379-1389.

28. Gong W., Wang Y., and Jia J. 2012. The effect of interacting downstream branches on saltwater intrusion in the Modaomen Estuary, China. Journal of Asian Earth Sciences, 45: 223-238.

29. Gosain A. K., Rao S., Srinivasan R., and Reddy N. G. 2005. Return-flow assessment for irrigation command in the Palleru River basin using SWAT model. Hydrological Processes, 19: 673-682.

30. Guedon, N. B., and Thomas, J. V. 2004. State of Mississippi Water Quality Assessment, section 305(b) Report 62. Mississippi Department of Environmental Quality, Jackson, MS.

31. Hamrick, J. M. 1996. A User's Manual for the Environmental Fluid Dynamics Computer Code (EFDC), The College of William and Mary, Virginia Institute for Marine Sciences, Special Report 331.

32. Hua L., Xiubin H., Yongping Y., and Hongwei N. 2012. Assessment of Runoff and Sediment Yields Using the AnnAGNPS Model in a Three-

Gorge Watershed of China. International Journal of Environmental Research and Public Health, 9:1887-1907.

33. Knisel W. G. 1980. CREAMS, a field scale model for chemicals, runoff and erosion from agricultural management systems. USDA Conservation Research Rept. No. 26, U.S. Department of Agriculture: Washington D. C.

34. Leonard R. A., W. G. Knisel, and D. A. Still. 1987. GLEAMS: Groundwater loading effects of agricultural management systems. Trans. of the ASAE, 30:1403-1418.

35. Licciardello F., D. A. Zema, S. M. Zimbone, and R. L. Bingner. 2007. Runoff and soil erosion evaluation by the AnnAGNPS model in a small Mediterranean watershed. Trans. of the ASAE, 50, 1585-1593.

36. Licciardello F., D. A. Zema, S. M. Zimbone, and R. L. Bingner. 2011. Runoff and soil erosion evaluation by the AnnAGNPS model in a small Mediterranean watershed. Trans. of the ASABE, 50(5): 1585-1593.

37. Mississippi Department of Environmental Quality (MDEQ), 2008. Sediment TMDL for the Yalobusha River Yazoo River Basin. PO Box 10385, Jackson, MS 39289-0385.

38. Mississippi Department of Environmental Quality (MDEQ). 2010. Total Daily Maximum Daily Load Program. Available at: http://www. deq.state.ms.us/MDEQ.nsf/page/TWB_Total_Maximum_Daily_Load_ Section?OpenDocument. Accessed on March 15, 2012.

39. Moriasi D. N., Arnold, J. G., Van Liew, M. W., Bingner, R. L., Harmel, R. D., and Veith, T. L. 2007. Model evaluation guidelines for systematic quantification of accuracy in watershed simulations. Transactions of the ASABE, 50(3): 885-900.

40. National Climatic Data Center (NCDC). 2012. Locate weather observation station record. Available at: http://www.ncdc.noaa.gov/oa/ climate/stationlocator.html. Accessed on April 14, 2012.

41. Neitsch S. L., Arnold, J. G., Kiniry, J. R., Williams, J. R., and King K. W. 2002. Soil and water assessment tool (SWAT), theoretical documentation, Blackland research center, grassland, soil and water research laboratory, agricultural research service: Temple, TX.

42. Neitsch S. L., Arnold J. G., Kiniry J. R., and Williams J. R. 2005. Soil and water assessment tool (SWAT), theoretical documentation, Blackland research center, grassland, soil and water research laboratory, agricultural research service: Temple, TX.

43. Nejadhashemi A. P., Woznicki S. A. and Douglas-Mankin K. R., 2011. Comparison of four models (STEPL, PLOAD, L-THIA, AND SWAT)

in simulating sediment, nitrogen, and phosphorus loads and pollutant source areas. Trans. of the ASABE, 54(3): 875-890.

44. Ouyang Y., J. Higman and J. Hatten. 2012. Estimation of dynamic load of mercury in a river with BASINS-HSPF model. Journal of Soils and Sediments, 12(2): 207-216.

45. Parajuli P. B., Mankin K. R. and Barnes P. L. 2008. Applicability of Targeting Vegetative Filter Strips to Abate Fecal Bacteria and Sediment Yield using SWAT. Agricultural Water Management, 95 (10): 1189-1200.

46. Parajuli P. B., Mankin K. R., and Barnes P. L. 2009. Source specific fecal bacteria modeling using soil and water assessment tool model. Bioresource Technology, 100 (2): 953-963.

47. Parajuli P. B. 2010a. Assessing sensitivity of hydrologic responses to climate change from forested watershed in Mississippi. Hydrological Processes, 24 (26): 3785-3797.

48. Parajuli P. B. 2010b. Methods for Modeling Livestock and Human Sources of Nutrients at Watershed Scale. MAFES Research Report, 24(8): 1-8.

49. Parajuli P. B. 2011. Effects of Spatial Heterogeneity on Hydrologic Responses at Watershed Scale. Journal of Environmental Hydrology, 19(18): 1-18.

50. Parajuli P. B. 2012. Evaluation of Spatial Variability on Hydrology and Nutrient Source Loads at Watershed Scale using a Modeling Approach. Hydrology Research, doi:10.2166/nh.2012.013.

51. Polyakov V., Fares A., Kubo D., Jacobi J., and Smith C. 2007. Evaluation of a non-point source pollution model, AnnAGNPS, in a tropical watershed. Environmental Modelling & Software, 22(11): 1617–1627.

52. Renard K. G., Foster G. R., Weesies G. A., McCool D. K., and Yoder D. C. 1997. Predicting Soil Erosion by Water: A Guide to Conservation Planning with the Revised Universal Soil Loss Equation (RUSLE). U.S. Department of Agriculture, Agriculture Handbook No. 703.

53. Rolle K., Gitau M. W., Chen G., and Chauhan A. 2012. Assessing fecal coliform fate and transport in a coastal watershed using HSPF. Water Science and Technology, 66:1096-1102.

54. Saad D. A., Schwarz G. E., Robertson D. M., and Booth N. L. 2011. A Multi-Agency Nutrient Dataset Used to Estimate Loads, Improve Monitoring Design, and Calibrate Regional Nutrient SPARROW Models. Journal of the American Water Resources Association, 47: 933-949.

55. Schwarz, G. E. 2008. A Preliminary SPARROW model of suspended sediment for the conterminous United States, U.S. Geological Survey Open-File Report 2008–1205, 8 p.

56. Sharpley A. N. and J. R. Williams. 1990. EPIC—Erosion/Productivity Impact Calculator: 1. Model Documentation. U.S. Department of Agriculture Technical Bulletin No. 1768.

57. Sheshukov A. Y., Siebenmorgen C. B., and Douglas-Mankin K. R. 2011. Seasonal and annual impacts of climate change on watershed response using an ensemble of global climate models. Trans. of the ASABE, 54(6): 2209-2218.

58. Shields F. D. Jr., Cooper C. M., Testa III, S., and Ursic, M. E. 2008. Nutrient Transport in the Yazoo River Basin, Research Report 60. U.S. Dept of Agriculture Agricultural Research Service National Sedimentation Laboratory: Oxford. http://www.ars.usda.gov / SP2UserFiles/person/5120/NSLReport60.pdf.

59. Shirmohammadi A., B. Ulen, L. F. Bergstrom, and W. G. Knisel. 1998. Simulation of nitrogen and phosphorus leaching in a structured soil using GLEAMS and a new sub-model. Trans. of the ASAE, 41(2): 353-360.

60. Sinha S., Liu X., and Garcia, M. H. 2012. Three-dimensional hydrodynamic modeling of the Chicago River, IL. Environmental Fluid Mechanics , Pages 1-24.

61. Smith R. A., Schwarz G. E., and Alexander R. B. 1997. Regional interpretation of water-quality monitoring data. Water Resources Research, 33 (12): 2781-2798.

62. Talei A., and Chua L. H. C. 2012. Influence of lag time on event-based rainfall-runoff modeling using the data driven approach. Journal of Hydrology, 438-439: 223-233.

63. U.S. Department of Agriculture (USDA). 2005. Soil data mart. Natural Resources Conservation Service. Available at: http://soildatamart.nrcs. usda.gov/Default.aspx. Accessed on May 23, 2012.

64. U.S. Department of Agriculture, National Agricultural Statistics Service (USDA/NASS). 2010. The cropland data layer. Available at: http://www. nass.usda.gov/research/Cropland/SARS1a.htm. Accessed on April 29, 2012.

65. U.S. Environmental Protection Agency (US/EPA). 2010. National summary of impaired waters and TMDL information. U.S. Environmental Protection Agency: Washington, D.C. Available at: http://iaspub.epa.gov/ waters10/attains_nation_cy.control?p_report_type=T. Accessed on June 28, 2012.

66. U.S. Geological Society (USGS). 1999. National elevation dataset. Available at: http://seamless.usgs.gov/index.php. Accessed April 29, 2012.

67. Vache K. B., Eilers J. M., and Santelmann M. V. 2002. Water quality modeling of alternative agricultural scenarios in the U.S. Corn Belt. Journal of American Water Resources Association, 38(3): 773-787.

68. Varanou E., Gkouvatsou E., Baltas E., and Mimikou M., 2002. Quantity and quality integrated catchment modeling under climate change with use of soil and water assessment tool model. ASCE Journal of Hydrological Engineering, 7(3): 228-244.

69. Vellidis G., P. Barnes, D. D. Bosch, and A. M. Cathey. 2012. Mathematical simulation tools for developing dissolved oxygen TMDLs. Trans of the ASABE, 49(4): 1003−1022.

70. Williams, J. R., C. A. Jones and P. T. Dyke. 1984. A modeling approach to determining the relationship between erosion and soil productivity. Trans. ASAE 27(1):129-144.

71. Williams J. R. and A. D. Nicks. 1985. SWRRB, a simulator for water resources in rural basins: an overview. In: D.G. DeCoursey (editor), Proc. of the Natural Resources Modeling Symp., Pingree Park, CO. USDA-ARS, ARS-30, pp. 17-22.

72. Yuan Y., Bingner R. L., and Rebich R. A. 2001. Evaluation of AnnAGNPS on Mississippi Delta MSEA watersheds. Trans. of the ASAE, 44(5): 1183−1190.

73. Yuan Y, Dabney S, and Bingner R. L. 2002. Cost/benefit analysis of agricultural BMPs for sediment reduction in the Mississippi Delta. Journal of Soil and Water Conservation, 57(5): 259−267.

74. Yuan Y., Locke M. A., and Bingner R. L. 2008. Annualized Agricultural Non-Point Source model application for Mississippi Delta Beasley Lake watershed conservation practices assessment. Journal of Soil and Water Conservation, 63(6): 542-551.

75. Yuan Y., R. L. Bingner M. A. Locke F. D. Theurer, and J. Stafford. 2011. Assessment of Subsurface Drainage Management Practices to Reduce Nitrogen Loadings Using AnnAGNPS. Applied Engineering in Agriculture, 27(3): 335-344.

76. Young R. A., Onstead C. A., Bosch D. D., and Anderson W. P. 1989. AGNPS: a nonpoint source pollution model for evaluating agricultural watersheds. Journal of Soil Water Conservation, 44(2):168−173.

77. Zhang G. H., B. Y. Liu and X. C. Zhang. 2008. Applicability of WEPP sediment transport equation to steep slopes. Trans of the ASABE, 51(5): 1675-1681.

78. Zuercher B. W., D. C. Flanagan, and G. C. Heathman. 2011. Evaluation of the AnnAGNPS model for atrazine prediction in Northeast Indiana. Trans of the ASABE, 54(3): 811-825.

Chapter 3

MODELLING HYDROLOGY OF A SINGLE BIORETENTION SYSTEM WITH HYDRUS-1D

Yingying Meng,[1,2] Huixiao Wang,[1] Jiangang Chen,[2] and Shuhan Zhang[2]

[1]College of Water Sciences, Beijing Normal University, Beijing 100875, China

[2]Beijing Water Science and Technology Institute, Beijing 100048, China

ABSTRACT

A study was carried out on the effectiveness of bioretention systems to abate stormwater using computer simulation. The hydrologic performance was simulated for two bioretention cells using HYDRUS-1D, and the simulation results were verified by field data of nearly four years. Using the validated model, the optimization of design parameters of rainfall return period, filter media depth and type, and surface area was discussed. And the annual hydrologic performance of bioretention systems was further analyzed under the optimized parameters. The study reveals that bioretention systems with underdrains and impervious boundaries do have some detention capability, while their total water retention capability is extremely limited. Better detention capability is noted for smaller rainfall events, deeper filter media, and design storms with a return period smaller than 2 years, and a cost-effective filter media depth is recommended in bioretention design. Better hydrologic effectiveness is achieved with a higher hydraulic conductivity and ratio of the bioretention surface area to the catchment area, and filter media whose conductivity is between the conductivity of loamy sand and sandy loam, and a surface area of 10% of the catchment area is recommended. In the long-term simulation, both infiltration volume and evapotranspiration are critical for the total rainfall treatment in bioretention systems.

INTRODUCTION

Rapid urbanization in watershed, with the increasing impervious area, implies both larger stormwater runoff volumes and peak flows and consequently reduces other components of the hydrologic cycle, for example, infiltration and evapotranspiration. Moreover, stormwater directly transports harmful substances from urban surfaces to downstream water systems, thus degrading the water quality. The negative impacts of urban stormwater have received widespread recognition [1], and maintaining stormwater quantity (e.g., flood peak and total volume) and quality (e.g., pollution) as close as the predevelopment levels has become increasingly popular. Bioretention, also known as rain garden, biofilter, or biofiltration, is a terrestrial-based water quantity and quality control practice that can be designed to mimic predevelopment hydrology (PGCo, 2007). It is thus commonly used as a source control technique to manage stormwater runoff in areas under urbanization and a retrofit technique in already developed areas [2]. Bioretention has also played an important role in the implementation of best management practice (BMP) and low impact development (LID) in America, water sensitive urban design (WSUD) in Australia, and sustainable urban drainage system (SUDS) in England.

There are many factors influencing the performance of bioretention systems, such as type of vegetation, depth of the filter media, size of the system relative to its catchment, and type of soil. Sizing, vegetation, construction technique, and soil mixture were all reported to have an important influence on the hydraulic conductivity of bioretention [3, 4]. The sizing of biofilters was also emphasized by Brown and Hunt III [5] who presented better reductions in runoff volume with deeper media depth. Furthermore, the hydraulic conductivity of the underlying soil and the internal water storage zone depth were also considered as primary factors influencing water reduction [6]. Overall, this research work on factors influencing the performance of bioretention was mainly based on column studies in laboratories or field studies [4, 5]. Because experimental observations were easily restricted by test conditions, unexpected results were sometimes reached. In the field study of Brown and Hunt III [5], for example, the surface storage volume of two bioretention cells was undersized because of design and construction errors, having substantial negative impacts on cell performance. Therefore, there is an increasing need to predict the hydrologic and water quality performance of bioretention systems using hydrologic model, which could be conveniently used in design, evaluation, or other purposes.

Initial model studies about bioretention did not include underdrains; for example, Heasom et al. [7] have attempted to predict the overflow volume in a bioinfiltration cell using one-dimensional hydrological model HEC-HMS. Considering an underdrain, He and Davis [8] developed a two-dimensional model simulating the subsurface flow. However, both models were based on individual rainfalls and were unable to perform continuous simulations and therefore they could not account for the changes in soil moisture conditions from previous rainfall events. The RECARGA model [9], widely used in the design and performance assessment of bioretention systems [10], allows for both continuous modelling and single-event modelling, but its minimum hourly rainfall interval makes it unable to conduct simulations for very short periods. Moreover, some parameters such as the number of underdrains and their depths and types of filter media could not be specified by the user, limiting the model's applications in some situations. As the water movement process in bioretention cells installed with underdrains is very similar to agricultural drainage pipes, modelling hydrologic performance in bioretention systems with DRAINMOD, an agricultural drainage model, has been common in recent years [11]. But DRANIMOD is unsuitable for conducting short-term simulations with a minimum calculation time of 1 month. Other models used in bioretention simulations involve SWMM (USEPA, 2010), SUSTAIN (USEPA, 2013), or MUSIC (eWater, 2013), but because of scale problem they are not appropriate for a single facility simulation.

One potential solution to reduce the frequent urban waterlogging disasters in Beijing in recent years is to control urban runoff at source as much as possible. As a source control technique, bioretention systems have advantages in ultraurban areas such as Beijing where land is unavailable for large control practices such as retention ponds, grassed swales, and constructed wetlands. The main objective of this study is to evaluate the hydrologic performance of bioretention facilities and provide instructive guidance for their design and application in Beijing. We developed a model tool to predict the hydrologic performance of a single bioretention facility and discuss the influence of different design parameters. Overall, the above-mentioned models have their own shortcomings in terms of scale, calculation time, configuration design, and other aspects. Comparatively, the HYDRUS-1D model [12] is a more appropriate model, with flexible water flow boundary conditions, a minimum calculation interval of 1 s, and unlimited simulation time. Hilten et al. [13] and Ladu et al. [14] both simulated the stormwater performance of a green roof using HYDRUS-1D. As a heterogeneous multilayer soil medium such as the green roof system, a bioretention system has the potential to be simulated by this model, while no data have yet been reported in the literature on the model's ability to model bioretention performance. In our study, the HYDRUS-1D model was devised based on input variables measured at two bioretention cells constructed in Beijing. The hydrology processes were measured at these facilities for nearly 4 years, and the collected data were used to calibrate and validate the model. The factors affecting the bioretention performance are discussed based on the simulation results for different design storms, filter media depths and types, and surface areas.

METHODS

Principle of the HYDRUS-1D Model

The hydrologic processes in bioretention systems consist of evapotranspiration, infiltration, and runoff generation. The water balance equation is given as follows:

$$ET = P - I - R \pm DSW, \tag{1}$$

where ET is the evapotranspiration, P is the precipitation, I is the infiltration, R is the runoff, and DSW is the change in soil water content. The total runoff R is given as follows:

$$R = RS + RB, \tag{2}$$

where RS is the surface runoff and RB is the bottom runoff from the drainage layer.

Field Study Site

Two parallel bioretention cells (Cells A and B) were constructed in Mentougou district in Beijing in 2010 (Figure 1). Each cell was 3 m * 2 m on the top surface, 2.2 m * 1.1 m on the bottom surface, and 1.1 m deep. Being designed to capture and treat stormwater runoff from a 60 m² impervious roof, each cell covers 10% of the catchment area. The cells were built with the following composition (Figure 2):

- A drainage layer at the base, containing a 110 mm diameter slotted PVC pipe (connected to the observation well for flow measurement) surrounded by 30 mm gravel with the diameter of 5–10 mm;

- Sixty centimeters of filter media: conventional media in Cell A, 97% of sand and soil, 3% of peat (both by volume); two layers in Cell B, one-third of sand and soil and two-thirds of blast furnace slag with the diameter of 5 mm on the top layer of 25 cm (both by volume), vermiculite on the bottom layer of 35 cm with the diameter of 0.5–5 mm to increase soil porosity [4];

- Five centimeters of mulch: shredded pine bark;

- Vegetation cover, with native plants of Ophiopogon japonicas and Iris tectorum for Cells A and B. Before choosing vegetation to conduct experiments on, we defined some criteria to ensure the vegetation suitable for bioretention construction. The criteria were defined to fit the climate in Beijing, specifically, drought, flood, pollution, salt, shade, and cold tolerant, with ornamental value locally selected, low cost, and low maintenance. Based on these requirements, we finally chose Ophiopogon japonicas and Iris tectorum;

- An overflow drain connected to the PVC pipe allowing a maximum ponding depth of 15 cm;
- An impervious geotextile on the sides and bottom to minimize migration of water into or out of the system.

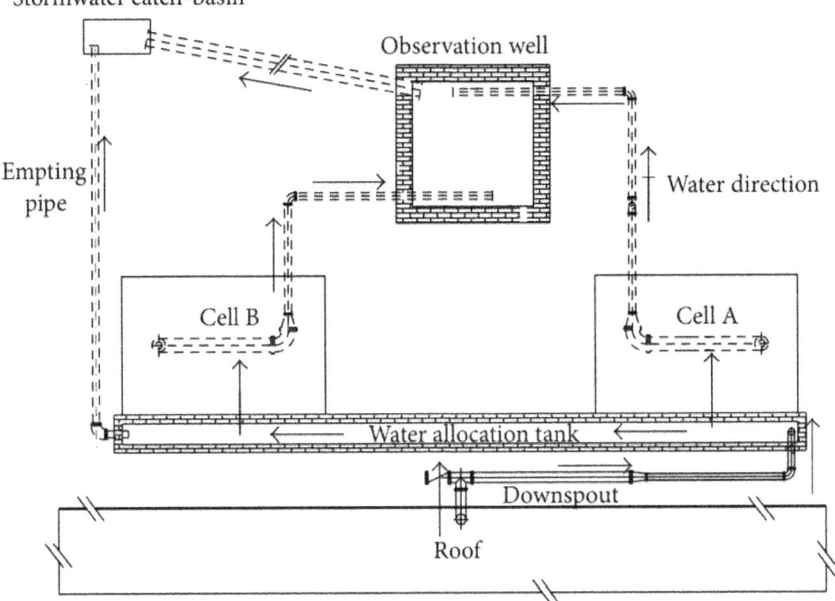

Figure 1: Plan of the bioretention study area (not to scale).

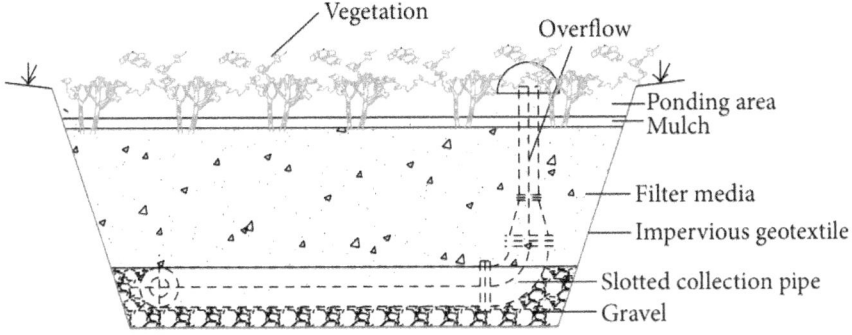

Figure 2: Schematic view of the bioretention cell.

Field Measurements

The infiltration and overflow volume which contributed to the bottom runoff both discharged through the PVC pipe, and its flow rate was measured using a 5 L measuring cup and a stopwatch every 5 min after the PVC pipe started to drain off water in the observation well, until there is no pipe flow. If ponding occurred, the water level in the ponding area was measured by a meter ruler every 5 min until ponding disappeared. These in situ data were collected during June to September in 2010, 2011, 2012, and 2013.

In the absence of natural rainfall, simulated stormwater was sometimes used for the experiments, using the reference method provided in Hsieh and Davis [15] to prepare the simulated stormwater. In artificial rainfall, the simulated stormwater with the theoretical volume from each cell's catchment area was mixed well and pumped into the cells evenly over an hour. The bottom runoff and ponding processes were also monitored.

Thirty-eight rainfall events were monitored, with thirty-three artificial events and five natural events. The artificial rainfall test results were mainly used for parameter calibration and validation of the HYDRUS-1D model, and the natural rainfall results were all used for model validation.

Modelling with HYDRUS-1D

Input requirements for HYDRUS-1D include geometry and time information, soil hydraulic and vegetation properties, initial and boundary conditions, and meteorological information, whose values are listed in Table1. The soil hydraulic parameters used in the van Genuchten model were measured using a high-speed centrifuge method (Table 2). Daily measurements of meteorological variables, including air temperature and humidity, atmospheric pressure, precipitation, wind speed and direction, and incoming shortwave and longwave radiation, were collected from the meteorological station in Beijing (number 54511) near the experimental site. The initial conditions are given in terms of water content, which is linearly distributed in the soil profile. The other parameters were specified according to the default values.

Table 1: The input information in the model for Cells A and B.

Input information	Parameters	Values	
		Cell A	Cell B
Initial condition	Minimum water content	0.36	0.34
	Maximum water content	0.38	0.36
	Maximum height at soil (cm)	15	15
Geometry information	Number of layers	1	2
	Depth of the soil (cm)	60	25, 35
Time information	Time duration (min)	300	120
	Time step (min)	0.00001~5	0.001~0.25
Water flow-soil hydraulic property model	Model	van Genuchten-Mualem	
Water flow boundary conditions	Upper boundary condition	Atmospheric BC with surface layer	
	Lower boundary condition	Seepage face	
Vegetation properties	Water uptake reduction model	Feddes	
	Crop height (cm)	30	60
	Root depth (cm)	20	30

Table 2: The soil hydraulic parameters for the van Genuchten model in different structure layers.

Structural layer	Residual water content (cm³/cm³)	Saturation water content (cm³/cm³)	α (min⁻¹)	(—)	Saturated hydraulic conductivity (cm/min)	(—)
Soil media in Cell A	0.065	0.410	0.025	1.69	0.083	0.5
Top media in Cell B	0.071	0.490	0.460	1.30	0.252	0.5
Bottom vermiculite in Cell B	0	0.448	0.002	1.44	0.518	0.5

In this study, the model was validated based on the measured bottom runoff and water level. Because the artificial and natural rainfalls were both individual events, we chose an artificial rainfall and a natural rainfall, respectively, as the rainfall events for validation. Statistics of the root-mean-square error (RMSE), mean relative error (MRE), and correlation coefficient (R^2) were then used to assess the accuracy of the simulation. The smaller the RMSE and MRE are, the more closely the R^2 approximated to zero and the better the model performed.

Model Application

Upon verifying the accuracy of the HYDRUS-1D model, simulations were run using different input variables to optimize the design of the bioretention systems. Four general scenarios with different rainfall return periods, filter media depths and types, and surface areas were investigated, and, through scenario analysis using the model, the influence of the different design parameters on the hydrologic performance of bioretention systems was studied. In the above simulations, storms were simulated as independent events. As HYDRUS-1D had no limitation on simulation time, the long-term hydrologic performance of bioretention systems was also assessed by inputting the annual meteorological data in 2012. The hydrologic performance of bioretention systems could be described in terms of the hydrologic effectiveness and the water detention and retention effects. Hydrologic effectiveness ((R_{hydro}) denotes the total rainfall runoff treated by the bioretention in whatever form, specified using (3). Water detention means water temporarily reserved by the system, which was demonstrated by bottom runoff delay (Δt_{dd}), bottom runoff peak flow delay (Δt_{pd}), and bottom runoff peak flow reduction (R_{pr}), given in (4)–(6). Comparatively, water retention (R_{reten}) refers to the water completely reserved by the system, given by (7). The ponding duration ((t_{pond}) can also reflect the hydrologic performance to some extent. Consider

$$R_{hydro} = \frac{V_{inflow}}{V_{runoff}} \times 100\%,$$

(3)

where V_{inflow} is the inflow volume of the bioretention cell and V_{runof} is the runoff volume from the catchment area. Consider

$$\Delta t_{dd} = t_{drain} - t_{inflow},$$

(4)

where t_{drain} is the time the bottom runoff appears and t_{inflow} is the time the inflow enters the bioretention cell. Consider

$$\Delta t_{pd} = t_{pdrain} - t_{pinflow},$$

(5)

where t_{pdrain} is the time the bottom runoff peak appears and $t_{pinflow}$ is the time the inflow peak appears. Consider

$$R_{pr} = \frac{q_{pinflow} - q_{pdrain}}{q_{pinflow}} \times 100\%,$$

(6)

where $q_{pinflow}$ is the inflow peak and q_{pdrain} is the bottom runoff peak. Consider

$$R_{reten} = \frac{V_{inflow} - V_{drain}}{V_{runoff}} \times 100\%,$$

(7)

where V_{drain} is the bottom runoff volume.

RESULTS AND DISCUSSION

Model Validation

In our experiments, overflow never occurred; thus, the bottom runoff volume was equal to the infiltration volume. Dividing the infiltration volume by the surface area of the bioretention cell, we obtain the infiltration rate. The comparison of the modelled infiltration rate by HYDRUS-1D and the observation values in Cells A and B are shown in Figure 3, with all of the simulated values in good agreement with observed ones. Because the filter media of Cell B had added blast furnace slag and vermiculite whose particle sizes were larger than soil particles, it was reasonable that the infiltration rate was higher in Cell B in both the artificial and the natural rainfalls, and the bottom runoff delay was obviously later in Cell A in the artificial rainfall.

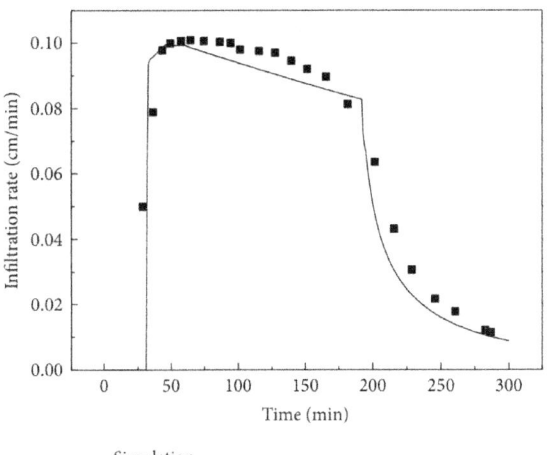

— Simulation
■ Observation

(a)

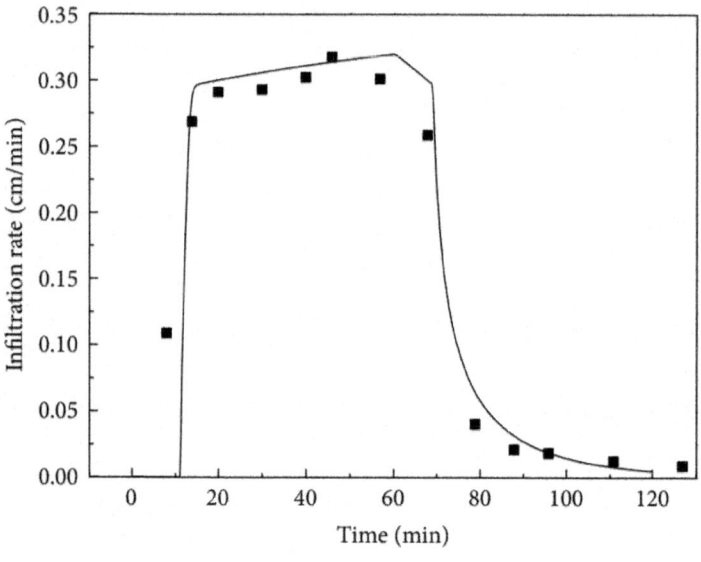

— Simulation
■ Observation

(b)

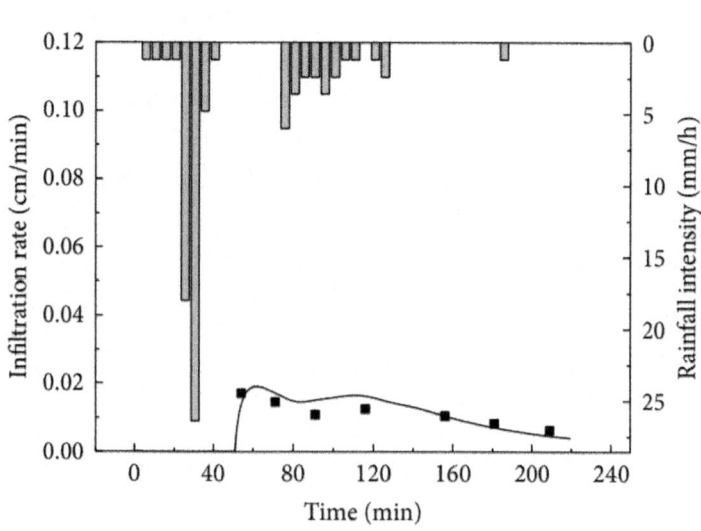

— Simulation
■ Observation
▨ Rainfall

(c)

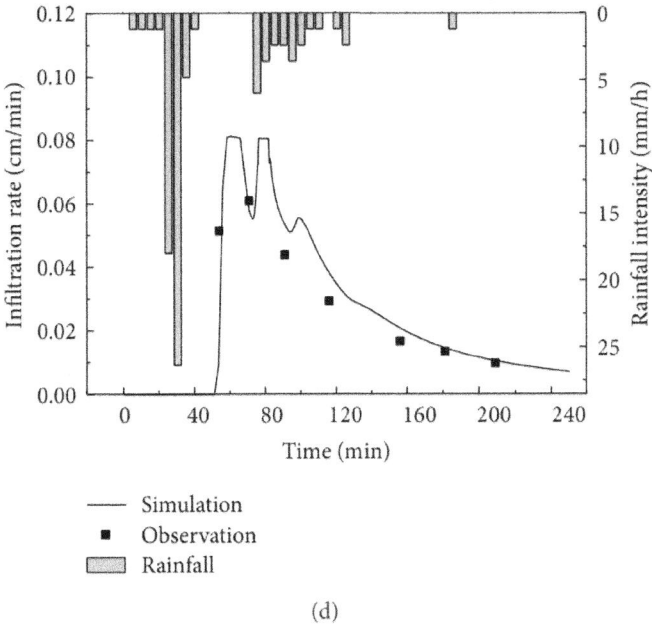

(d)

Figure 3: Observed and simulated infiltration rate in the artificial rainfall for validation in (a) Cell A and (b) Cell B and in the natural rainfall for validation in (c) Cell A and (d) Cell B.

The statistics for the RMSE, MRE, and R^2 between the simulated and observed values are shown in Table 3. The evaluation results were acceptable with the RMSEs all approximating zero, the MRE almost at 0.20, a higher R^2 (>0.9) in the artificial rainfall, and a relatively smaller R^2 (>0.6) in the natural rainfall, which might be the result of fewer observed data.

Table 3: Statistics for simulation accuracy assessment.

Cell ID	RMSE		MRE			
	Artificial rainfall	Natural rainfall	Artificial rainfall	Natural rainfall	Artificial rainfall	Natural rainfall
A	0.014	0.003	0.16	0.22	0.95	0.76
B	0.036	0.017	0.23	0.25	0.97	0.61

For the ponding process, the simulated and measured water levels in Cell A in the artificial rainfall for validation are presented in Figure 4. The observed and calculated values were well matched, as the RMSE, MRE, and R^2 were 0.34, 0.059, and 0.99, respectively. Flooding never occurred in Cell B in the experiments, while a thin layer of water (less than 2 cm) appeared in the simulation results, probably because the water retention of mulch is not taken into account by the model.

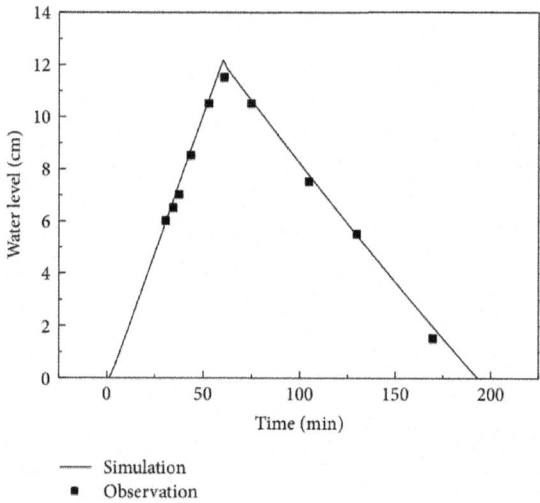

Figure 4: Observed and simulated water level in the artificial rainfall for validation in Cell A.

Overall, the HYDRUS-1D model was capable of capturing the hydrologic processes in bioretention systems with reasonable accuracy and can thus be used to assess bioretention performance under different circumstances.

Design Parameters Optimization

Rainfall Return Periods

Twenty-four-hour design storms with different return periods in the central area of Beijing were calculated according to the Hydrologic Manual of Beijing [16]. With the other input requirements the same as those in the Cell A simulation, the simulated results for different design storms are presented in Figure 5. From Figure 5, it is clear that the hydrologic effectiveness (R_{hydro}) decreased with the rainfall return period because of the increased total runoff volume from the catchment area. Thus, in rainfalls with return periods greater than 5 years, the peak flow reduction (R_{pr}) was higher because R_{hydro} was lower, and only in rainfalls with return periods smaller than 2 years did R_{pr} really increase because of the higher R_{hydro}. Therefore, greater hydrologic effectiveness is the premise for better water treatment. The water detention effect can be further evaluated using the bottom runoff delay (Δt_{dd}) and bottom runoff peak flow delay (Δt_{pd}) besides R_{pr}, and Δt_{dd} gradually decreased when the rainfall return period was greater than 5 years, while an obvious decrease was observed with a rainfall return period smaller than 5 years. However, Δt_{pd} changed dramatically with the rainfall return period, with a sharp increase for a return period smaller than 5 years, a significant decrease for a return period between 5 and 10 years, and a gentle decrease for a return period greater than 10 years. This is because of the two peaks of the 24-hour design storm in Beijing, and the bioretention could only abate the first peak with a rainfall return period greater than 10 years, while, in rainfalls with a smaller return period, it successfully eliminated the first peak, resulting in the delayed bottom runoff peak caused by the second rainfall peak. Furthermore, the variation in ponding duration (t_{pond}) was very similar to Δt_{dd}, with a gradual increase in rainfalls for a return period greater than 5 years, while there was a significant increase in rainfalls with a return period smaller than 5 years. It should be noted that to prevent mosquito breeding and maintain vegetation growth ponding duration is required to be shorter than 48 h (PGCo, 2007), and the simulated t_{pond} between 4 and 20 h under different design storms all met the requirement. In summary, the water detention effect in bioretention systems is much better in rainfalls with smaller return periods. Referring to the total water retention effect, the decrease in water retention (R_{reten}) with the rainfall return period was not obvious, and R_{reten} was smaller than 10% even with the smallest storm return period of 1 year, showing that bioretention measures could store very limited stormwater, which might be because of the completely impervious sides and bottom surface in our study site.

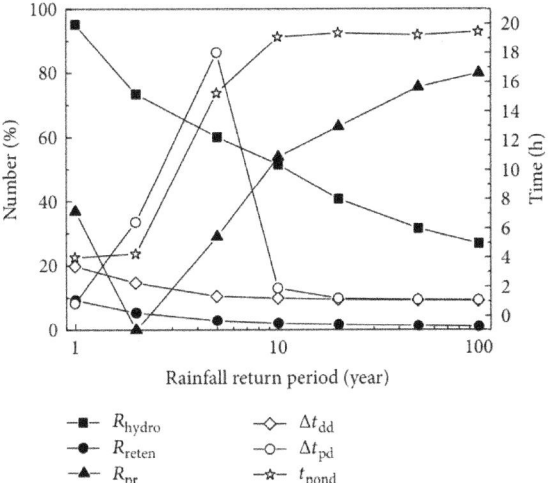

Figure 5: Hydrologic performance of bioretention measures under rainfalls with different return periods.

Conclusively, with impervious surroundings, bioretention facilities only regulate the inflow runoff discharge process, their total water retention effect is very limited, and their detention effect is significantly better in small rainfalls with return periods smaller than 2 years. This agrees with the research results of Davis [17] and Li et al. [18], who indicated that the greatest impact was noted for the smaller events. Thus, a 2-year storm or less is recommended for use as the design storm for bioretention measures. Therefore, the following discussion on the design parameters is based on a 1-year storm.

Filter Media Depths

Taking into account the demand for water quality improvement [19] and the underdrains successfully connected with the municipal storm sewer, the input filter media depths were from 30 to 90 cm. The variation in hydrologic performance is presented in Figure 6. The bottom runoff peak flow reduction (R_{pr}) and bottom runoff delay (Δt_{dd}) increased significantly with the filter media depth, while the total water treated (R_{hydro}), water retention (R_{reten}), bottom runoff peak flow delay (Δt_{pd}), and ponding duration (t_{pond}) showed almost no change.

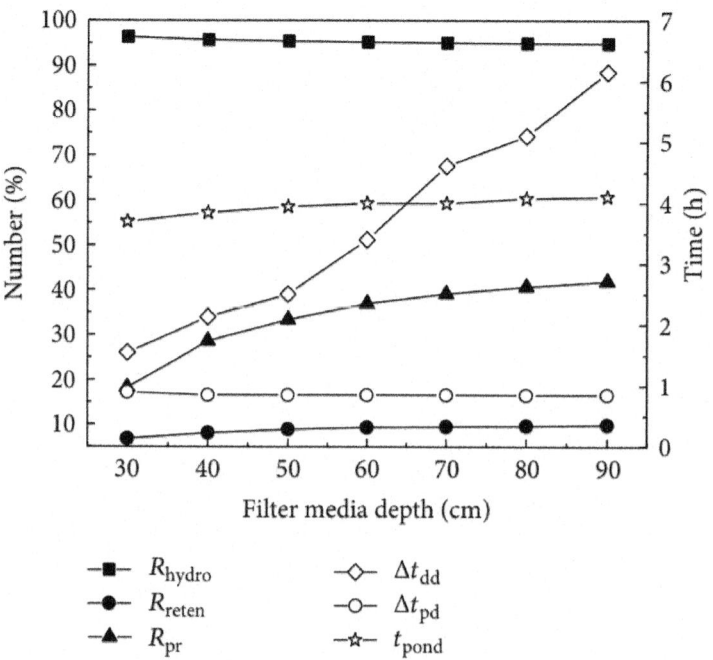

Figure 6: Hydrologic performance of bioretention measures under different filter media depths.

This indicated that the filter media depth only has a beneficial impact on water detention, with better water detention in deeper filter media, while it has little or no influence on hydrologic effectiveness, water retention, and ponding control, which may also be attributed to the impervious surroundings. As in the study of Brown and Hunt III [5], the exfiltration volume in bioretention systems without the impermeable membrane is much higher in the deeper media cells because of greater storage volume in the media and more exposure to the side walls, leading to the better reductions in runoff volume.

As one of the main costs in constructing bioretention cells, the filter media depth should not be too great; thus, a cost-effective depth should be used in bioretention design. This could also refer to the media depth requirements in North Carolina, where vegetation is the determined factor and the minimum media depth is 0.6 m for cells vegetated with grass or shallow rooted plants and 0.9 m for cells vegetated with shrubs or trees (NCDENR, 2009).

Types of Filter Media

Nine soils, from sand to clay with declining saturated hydraulic conductivity, included in the soil catalog of the HYDRUS-1D model, were used as input filter media, and in each case the hydrologic performance was simulated with default soil hydraulic parameters in the model (Figure 7). It was obvious that filter soil played an important role in hydrologic effectiveness and the total water treated by the system (R_{hydro}) decreased with declining hydraulic conductivity from sand to clay. This agreed with the research results of Le Coustumer et al. [2] who even proposed that the initial specified hydraulic conductivity of filter media was the critical determinant of the long-term hydraulic behavior of a biofilter. In this study, with a high infiltration rate, the sandy texture soils treated all the rainfall runoff without ponding, while with low infiltration rate the clay soil only treated 23.6% of the total rainfall runoff while it overflowed the other runoff and was always ponding. Without higher hydrologic effectiveness, the obvious increase in the water detention parameters of bottom runoff peak reduction (R_{pr}), bottom runoff delay (Δt_{dd}), and bottom runoff peak flow delay (Δt_{pd}) with the declining infiltration rate makes no sense. Moreover, the slight increase in R_{reten} with declining hydraulic conductivity was also attributed to the poor retention effect of the whole bioretention system, which has already been mentioned above.

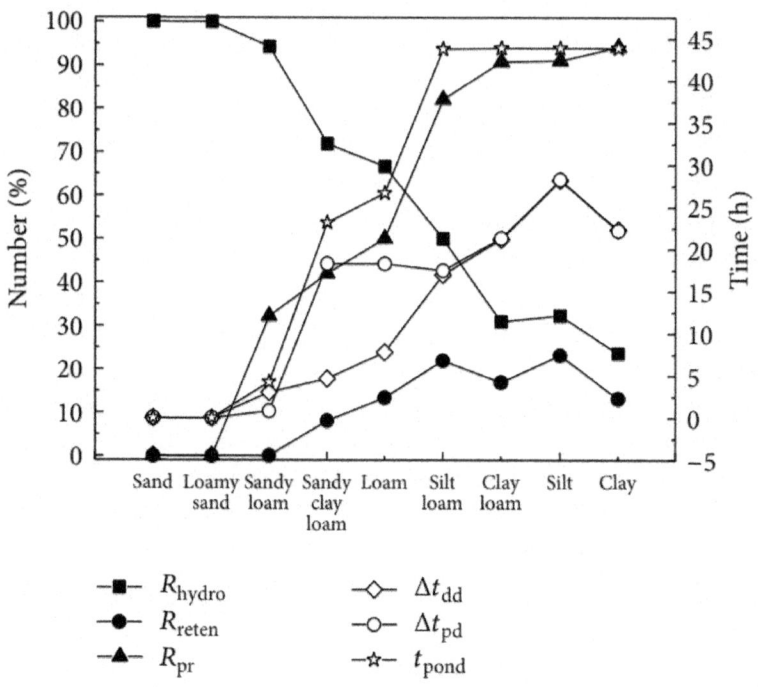

Figure 7: Hydrologic performance of bioretention measures under different filter soils.

Recommendations for hydraulic conductivity of soil media vary from one country to another [2], with at least 12.5 mm/h in New Zealand and America, between 36 and 360 mm/h in Austria and between 50 and 200 mm/h in Australia. The saturated hydraulic conductivity of soils included in the soil catalog of the HYDRUS-1D model, taken from Carsel and Parrish [20], is shown in Table 4. Considering the conductivity requirement in different countries and its potential reduction with time, it is recommended that soils between loamy sand and sandy loam are used as the filter media in bioretention design enabling high hydrologic effectiveness (R_{hydro}) and medium detention and retention effects (R_{pr}, Δt_{dd}, Δt_{pd}, and). Moreover, the ponding duration (t_{pond}) in the simulations of loamy sand and sandy loam was shorter than 5 h, also maintaining better vegetation growth. In USEPA (1999), the sandy loam soil has already been recommended to be used in bioretention systems. In our experiment site, the saturated hydraulic conductivity of soil media in Cells A and B was, respectively, 49.6 and 151 mm/h (0.083 and 0.252 cm/min in Table 2), very close to the infiltration performance of sandy loam and loamy sand; thus, the two facilities both perform well in terms of water infiltration.

Table 4: The saturated hydraulic conductivity of soils included in the soil catalog of the HYDRUS-1D model.

Soil type	Sand	Loamy sand	Sandy loam	Sandy clay loam	Loam	Silt loam	Clay loam	Silt	Clay
Saturated hydraulic conductivity (mm/h)	297	146	44	13.1	10.4	4.5	2.6	2.5	2

Surface Areas

General design guidelines suggest that the bioretention basin is approximately 5–7% of the effective upslope drainage area contributing to runoff (USEPA, 1999). With a surface area of 1–100% of the catchment area, the hydrologic performance of the bioretention is presented in Figure 8. As the surface area increased, the hydrologic effectiveness (R_{hydro}) and water retention effect (R_{reten}) both increased; meanwhile, the ponding duration (t_{pond}) decreased. This agreed with the research results of Jones and Hunt [21] who suggested that large bioretention areas could reduce surface ponding times. However, the three hydrologic parameters for the water detention effect, bottom runoff delay (Δt_{dd}), bottom runoff peak flow delay (Δt_{pd}), and bottom runoff peak flow reduction (R_{pr}), behaved differently with Δt_{dd} increasing with the surface area, while Δt_{pd} and R_{pr} changed drastically. This may be a function of the variation in hydrologic effectiveness, and, with a quick increase in R_{hydro} between area ratios of 1% and 10%, both Δt_{pd} and R_{pr} changed irregularly. When R_{hydro} approached 100% with an area ratio larger than 10%, Δt_{pd} increased gradually, but R_{pr} still had irregular variations, probably because of the combined effects of different inflow volumes and the constant hydraulic capacity of the filter media. However, it was evident that larger surface areas achieved better hydrologic performance, which was also confirmed by Le Coustumer et al. [2] who found that a larger surface area compensated for low conductivity by providing a greater filter area and ponding volume. From another perspective, Dussaillant et al. [22] showed that bioretention with an area of 10–20% of the contributing impervious area maximized groundwater recharge. However, in reality, as the bioretention area increases, the value of land increases especially in current Chinese cities and the facility becomes more costly. Considering the cost of land, a cost-effective surface area is recommended in bioretention design. From Figure 8, when the bioretention covers more than 10% of the catchment area, the total hydrologic effectiveness tends towards stability, and the water detention and retention effects change for the better; moreover, a ponding duration shorter than 4 h is also acceptable, and thus the surface area of 10% of the catchment area may be a reasonable compromise.

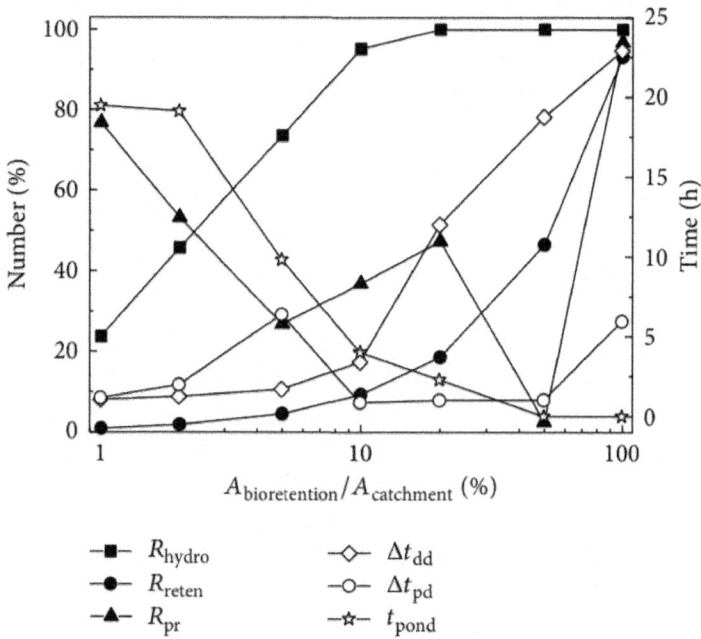

Figure 8: Hydrologic performance of bioretention measures under different bioretention areas of the catchment area.

Long-Term Hydrologic Performance

Using a medium filter depth of 60 cm, the filter type used in Cell A, which approximated the recommended sandy loam soil, and the recommended surface area of 10% of the catchment area, the long-term hydrologic performance of bioretention systems was assessed by inputting the annual meteorological data in 2012. The water volumes variation results are given in Figure 9. Because of the impervious sides and bottom surface of the bioretention system, it could be expected that the infiltration volume would take a large share of the total rainfall and the soil retention water may occupy a much lower percentage, which is proved in Figure 9 with the infiltration volume increasing rapidly with time, while the soil retention volume always fluctuates in the year. Meanwhile, the vegetation transpiration volume also increases evidently with time, which showed that, in the long run, evapotranspiration played an important role in the hydrology efficiency of bioretention systems. This was also confirmed by Dussaillant et al. [22], who reported that plant evapotranspiration during interstorm periods provided a greater available soil water storage capacity for the next rainfall event. It could be seen in Figure 9 that, after a year of operation, the infiltration, evaporation, transpiration, soil retention, and overflow volumes in the bioretention system were 560 mm, 6.3 mm, 146 mm, 1.4 mm, and 20 mm, respectively, contributing to 75.7%, 0.9%, 19.7%, 0.2%, and 2.7% of the total rainfall in 2012.

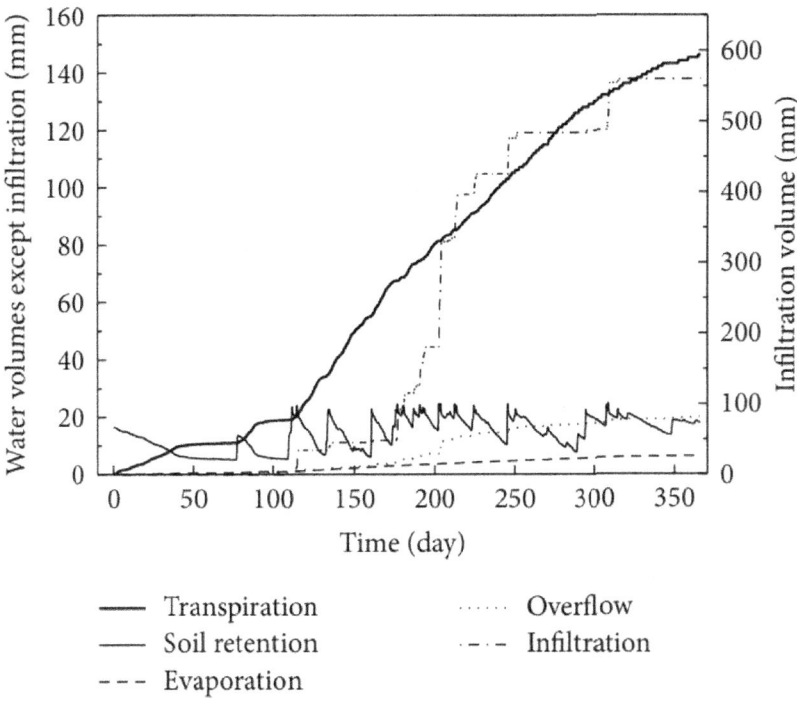

Figure 9: Accumulative water volumes variation with time in the long-term simulation using the annual meteorological data in 2012.

Furthermore, some researchers provided that plants improved filter performance; for example, Archer et al. [23] reported that root growth increased hydraulic conductivity as a result of macropores created by root dieback, and Le Coustumer et al. [4] showed that plants with thick roots maintained system permeability over time compared with plants with finer roots. However, plants only influence evapotranspiration through their growth characteristics of height and root depth in HYDRUS-1D, and, as shown in Figure 10, evapotranspiration volume increased evidently with plant height. Presently, the model is unable to simulate the effect of plants on the permeability of the system.

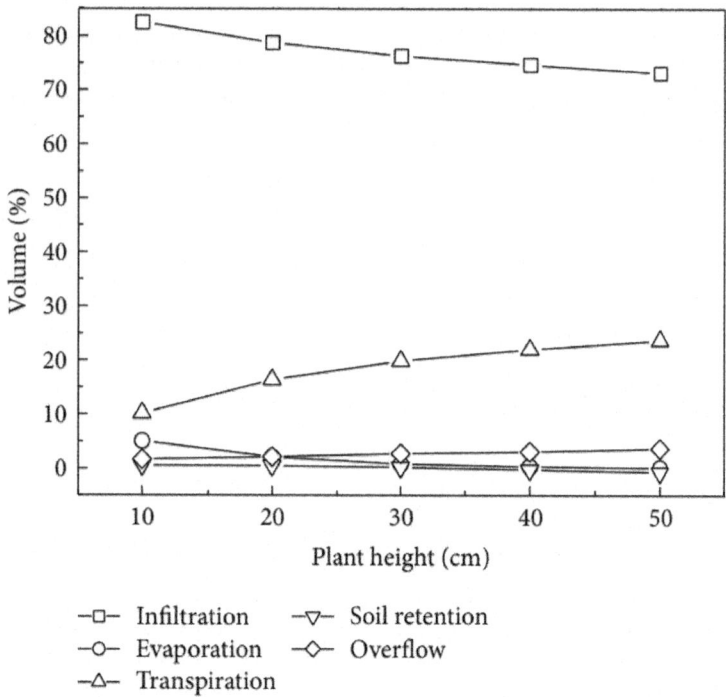

Figure 10: Water volume variation in bioretention systems with different plant heights.

CONCLUSIONS

Because stormwater management in urbanized areas has become ubiquitous, bioretention systems have been introduced as an effective source control technique to reduce runoff from impervious surfaces.

In this study, the hydrologic performance of two bioretention cells is modelled using HYDRUS-1D, with simulation results verified by field data. In the study, HYDRUS-1D accurately predicted infiltration and ponding processes in the bioretention cells.

The influence of different design parameters on the rainfall return period, media depth and type, and surface area to hydrologic performance was evaluated using the calibrated HYDRUS-1D model. It was shown that bioretention systems with underdrains and impervious boundaries have only some detention effect on bottom runoff delay, bottom runoff peak flow delay, and bottom runoff peak flow reduction, and their total water retention effect was very limited. Better detention effect was noted for smaller rainfall events, and a 2-year or less design storm was consequently recommended.

Filter media depth also had a significant impact on water detention but little or no effect on the total water treated. Better water detention appeared in deeper filter media, while, considering the filter cost, a cost-effective depth was recommended in bioretention design.

Both the hydraulic conductivity of filter media and surface area size influenced hydrologic effectiveness greatly, and better hydrologic effectiveness was reached with higher hydraulic conductivity and surface area ratio of the catchment area. Filter media with conductivity between loamy sand and sandy loam was recommended in bioretention design, enabling some conductive and retention effect as well as vegetation growth. Considering the cost of land, the cost-effective surface area was recommended in bioretention design, and the surface area of 10% of the catchment area may be a reasonable compromise.

Using the optimized design parameters for the rainfall return period, filter media depth and type, and surface area size, the long-term hydrologic performance of bioretention systems was further evaluated. As expected, the runoff inflow into the bioretention cell was mainly attenuated via infiltration, while at the same time evapotranspiration played an important role in the long run, contributing to 20.6% of the total rainfall in 2012.

Filter media play a very important role in hydrologic performance of bioretention measures, as conductivity and water retention capacity directly affect the infiltration, storage, and pollutant removal of inflow runoff. The pollutant transport process through the bioretention was not included in this study; thus, simulations of water quality improvement performance could be tried using the solute transport function of the HYDRUS-1D model, and, on this basis, the potential design parameters for better pollutant removal could be discussed, providing more references for the promotion and application of bioretention measures.

ACKNOWLEDGMENTS

This work is supported by the National Science Foundation of China under Grant no. NSFC-51179009 and National Major Science and Technology Programs under Grant no. 2013ZX07304-001.

REFERENCES

1.   B. E. Hatt, N. Siriwardene, A. Deletic, and T. D. Fletcher, "Filter media for stormwater treatment and recycling: the influence of hydraulic properties of flow on pollutant removal," Water Science and Technology, vol. 54, no. 6-7, pp. 263–271, 2006.

2. S. le Coustumer, T. D. Fletcher, A. Deletic, S. Barraud, and J. F. Lewis, "Hydraulic performance of biofilter systems for stormwater management: Influences of design and operation," Journal of Hydrology, vol. 376, no. 1-2, pp. 16–23, 2009.

3. D. D. Carpenter and L. Hallam, "Influence of planting soil mix characteristics on bioretention cell design and performance," Journal of Hydrologic Engineering, vol. 15, no. 6, pp. 404–416, 2010.

4. S. Le Coustumer, T. D. Fletcher, A. Deletic, S. Barraud, and P. Poelsma, "The influence of design parameters on clogging of stormwater biofilters: a large-scale column study," Water Research, vol. 46, no. 20, pp. 6743–6752, 2012.

5. R. A. Brown and W. F. Hunt, "Impacts of media depth on effluent water quality and hydrologic performance of undersized bioretention cells," Journal of Irrigation and Drainage Engineering, vol. 137, no. 3, pp. 132–143, 2011.

6. R. A. Brown and W. F. Hunt, "Underdrain configuration to enhance bioretention exfiltration to reduce pollutant loads," Journal of Environmental Engineering, vol. 137, no. 11, pp. 1082–1091, 2011.

7. W. Heasom, R. G. Traver, and A. Welker, "Hydrologic modeling of a bioinfiltration best management practice," Journal of the American Water Resources Association, vol. 42, no. 5, pp. 1329–1347, 2006.

8. Z. He and A. P. Davis, "Process modeling of storm-water flow in a bioretention cell," Journal of Irrigation and Drainage Engineering, vol. 137, no. 3, pp. 121–131, 2011.

9. A. R. Dussaillant, A. Cuevas, and K. W. Potter, "Raingardens for stormwater infiltration and focused groundwater recharge: simulations for different world climates," Water Science and Technology: Water Supply, vol. 5, no. 3-4, pp. 173–179, 2005.

10. T. M. Muthanna, M. Viklander, and T. T. Thorolfsson, "An evaluation of applying existing bioretention sizing methods to cold climates with snow storage conditions," Water Science and Technology, vol. 56, no. 10, pp. 73–81, 2007.

11. R. A. Brown, W. F. Hunt, and R. W. Skaggs, "Modeling bioretention hydrology with DRAINMOD,"Low Impact Development, pp. 441–450, 2010.

12. J. Šimůnek, M. Šejna, H. Saito, M. Sakai, and M. T. Van Genuchten, "The HYDRUS-1D software package for simulating the movement of water, heat, and multiple solutes in variably saturated media, Version 4.0," HYDRUS software series, 3: 315, 2008.

13. R. N. Hilten, T. M. Lawrence, and E. W. Tollner, "Modeling stormwater runoff from green roofs with HYDRUS-1D," Journal of Hydrology, vol. 358, no. 3-4, pp. 288–293, 2008.

14. J. L. C. Ladu, P. L. Demetry, T. O. Henry, and X. Lijun, Modeling Storm Water Runoff from Green Roofs with HYDRUS-1D, 2010.

15. C. Hsieh and A. P. Davis, "Evaluation and optimization of bioretention media for treatment of urban storm water runoff," Journal of Environmental Engineering, vol. 131, no. 11, pp. 1521–1531, 2005.

16. Water Bureau of Beijing, Hydrologic Manual of Beijing, Water Bureau of Beijing, Beijing, China, 1999.

17. A. P. Davis, "Field performance of bioretention: hydrology impacts," Journal of Hydrologic Engineering, vol. 13, no. 2, pp. 90–95, 2008.

18. H. Li, L. J. Sharkey, W. F. Hunt, and A. P. Davis, "Mitigation of impervious surface hydrology using bioretention in North Carolina and Maryland," Journal of Hydrologic Engineering, vol. 14, no. 4, pp. 407–415, 2009.

19. H. Li and A. P. Davis, "Urban particle capture in bioretention media. I: laboratory and field studies,"Journal of Environmental Engineering, vol. 134, no. 6, pp. 409–418, 2008.

20. R. F. Carsel and R. S. Parrish, "Developing joint probability distributions of soil water retention characteristics," Water Resources Research, vol. 24, no. 5, pp. 755–769, 1988.

21. M. Jones and W. F. Hunt, "Effect of bioretention on runoff temperature in trout sensitive regions," inProceedings of the World Environmental and Water Resources Congress, pp. 1582–1588, May 2009.

22. A. Dussaillant, K. Cozzetto, K. Brander, and K. Potter, "Green-Ampt model of a rain garden and comparison to Richards equation model," Sustainable World, vol. 6, pp. 891–900, 2003.

23. N. A. L. Archer, J. N. Quinton, and T. M. Hess, "Below-ground relationships of soil texture, roots and hydraulic conductivity in two-phase mosaic vegetation in South-East Spain," Journal of Arid Environments, vol. 52, no. 4, pp. 535–553, 2002.

Chapter 4

THE REVIEW OF GRACE DATA APPLICATIONS IN TERRESTRIAL HYDROLOGY MONITORING

Dong Jiang,[1] Jianhua Wang,[2] Yaohuan Huang,[1] Kang Zhou,[3] Xiangyi Ding,[2] andJingying Fu[1]

[1]Institute of Geographical Sciences and Natural Resources Research, Chinese Academy of Sciences, Beijing 100101, China

[2]State Key Laboratory of Simulation and Regulation of Water Cycle in River Basin, Department of Water Resources, China Institute of Hydropower & Water Resources Research, Beijing 100038, China

[3]Computer Network Information Center, Chinese Academy of Sciences, Beijing 100190, China

ABSTRACT

The Gravity Recovery and Climate Experiment (GRACE) satellite provides a new method for terrestrial hydrology research, which can be used for improving the monitoring result of the spatial and temporal changes of water cycle at large scale quickly. The paper presents a review of recent applications of GRACE data in terrestrial hydrology monitoring. Firstly, the scientific GRACE dataset is briefly introduced. Recently main applications of GRACE data in terrestrial hydrological monitoring at large scale, including terrestrial water storage change evaluation, hydrological components of groundwater and evapotranspiration (ET) retrieving, droughts analysis, and glacier response of global change, are described. Both advantages and limitations of GRACE data applications are then discussed. Recommendations for further research of the terrestrial water monitoring based on GRACE data are also proposed.

INTRODUCTION

Intensified impact of human activities and global change inevitably lead to the changes in the water cycle, including the spatial and temporal distribution and the total amount of water resources [1, 2]. Terrestrial hydrology, which is an important indicator in global change, affects the global climate system in energy, water, and biogeochemical [3]. Traditional monitoring of terrestrial hydrology mainly depends on site measurements or model simulations, which is costly and time-consuming. Furthermore, for the scale effects of the conversion from observation points to large region, additional error will be brought into spatially continuous data obtained by statistical interpolation methods. The accuracy of interpolation results may significantly decrease in areas far away from observation sites. Hydrological models and land surface models solve the problem of simulation of terrestrial hydrology on the plane scale to some degree. However, the lack of systematic global fine-scale hydrological data increases the model uncertainty and lowers the simulation accuracy. This restricts the practical application of hydrological models and land surface models in the monitoring of terrestrial hydrology. With the rapid development of remote sensing technology, remote sensing holds significant potential to the regional hydrological researches. By offering a useful and cost-effective approach to rapidly monitor terrestrial hydrological parameters, remote sensing technology has been widely used in hydrology. Recently, combination models of remote sensing technology, observation network, and hydrological models have been common in terrestrial hydrology studies. However, imaging satellite techniques and satellite altimetry can only be used in the monitoring of surface water that represents just one component of total water storage. They can be hardly applied in the monitoring of other components of terrestrial water such as soil water and groundwater, which impeded further applications of these technologies in terrestrial hydrology.

Launched in March 2002, the Gravity Recovery and Climate Experiment (GRACE) can provide an operational product in the form of global gravity fields every ~30 days [4]. After several gravity effects have been released (e.g., atmospheric mass variations and ocean tides), precise observations of the time-series global gravity field changes make the spatial monitoring of water storage changes at large scale such as basins possible. Through 10 years of development, GRACE satellite data processing and corresponding TWS retrieval algorithms are improved continuously. They are able to detect the changes of TWS within the accuracy of 1.5 cm on a wide range of spatial scales and seasonal time scales [5]. Compared to imagery remote sensing methods, they provide alternative technical methods for the estimation of glaciers, surface snow, soil water, surface water, groundwater, evaporation, and other components in the terrestrial hydrological system. It provides a new technique for the large-scale monitoring of terrestrial water cycle. The objective of this paper is to present an overview of GRACE data applications in terrestrial hydrology monitoring, including GRACE data processing methods and TWS change retrieval and their application in terrestrial hydrology and related fields.

GRACE DATA

GRACE gravity satellite program was jointly developed by the National Aeronautics and Space Administration (NASA) of the United States and the German Aerospace Center (DLR) with the objective of providing spatiotemporal variations of the Earth's gravity field. The GRACE satellite program can be also used to detect the atmosphere and ionosphere environment. The US Jet Propulsion Laboratory (JPL) is responsible for the project management of the GRACE gravity satellite program. Monthly gravity field solutions are computed at the University of Texas at Austin Center for Space Research (CSR), the German Research Centre for Geosciences Potsdam (GFZ), JPL, Groupe de Recherche de Geodesie Spatiale (GRGS), and the Delft Institute of Earth Observation and Space Systems (DEOS) as well as Delft University of Technology, among others [4,6–8]. GRACE utilizes a state-of-the-art technique to trace the spatiotemporal gravity field with an increased sensitivity by tracking the micrometer-precise intersatellite range and range-rate observations between two coplanar, low altitude (300 km~500 km) satellites and the distance between which is about 220 km. To precisely measure the distance changes between two satellites at micron meter level accuracy, a K-band ranging (KBR) system based on carrier phase measurements in the K (26 GHz) and Ka (32 GHz) frequencies is provided. Besides, four key instrumentations of a GPS receiver (space-proofed multichannel, two-frequency), capacitive accelerometer, laser retroreflector (LRR), and star camera are equipped on board of each GRACE satellite [4, 7].

By analyzing the relationship between orbits and forces of GRACE satellites, the Earth's gravity field variety is estimated based on the dynamic equations of satellites motion. It monitors the time-varying characteristics of long-wavelength gravity field on a 15~30-day or longer time scale. Large-scale mass redistribution (mass distribution changes over time) in the Earth system reflects the interaction between substances of various forms in the Earth's internal system (atmosphere, oceans and solid crust, viscous mantle, liquid outer core, and solid inner core), which are the important subjects of Earth sciences. Compared to the average Earth's gravity field, the time-varying quantity of the gravity field is very small, but it contains important geophysical information. It reveals the movement, distribution, and changes of all substances in the Earth system and reflects the interaction between the atmosphere, terrestrial water, oceans, and the solid Earth [6].

The time variation of global gravity field caused by the influence of solid Earth (including the inner core and the outer core) is mainly manifested on a 10-year or a longer time scale. However, temporal variation of gravity is mainly caused by the redistribution influence of atmosphere, oceans, and water storages in the surface fluid envelopes of the earth on a seasonal or interannual time scale [9, 10]. The gravity effects of tides (solid Earth, oceans, and atmosphere) and nontidal (atmosphere and oceans) are reduced in the data processing of GRACE gravity field model. Therefore, by excluding the errors in gravity field model and the atmosphere and oceans models, GRACE time-varying gravity field reflects the nonatmospheric and nonocean mass variations due to water mass variations on the continental area. On a seasonal or shorter time scale, it provides information on changes in terrestrial water storage in large river basin [11].

HYDROLOGICAL APPLICATIONS

TWS Changes Monitoring

The TWS is the most direct hydrological parameter obtained by GRACE monitoring. For the main goal of GRACE satellite is to monitor the Earth's gravity field variations, early researches are focused on the feasibility and accuracy of TWS retrieval from GRACE data. Wahr et al. [11] pointed out that the GRACE-based TWS retrieval required the consideration of the impacts from short-wave noises and leakage errors, and the TWS retrieval result is valid for large-scale river basins. Further research of Rodell and Famiglietti showed that TWS accuracy could be improved by increasing the monitoring temporal interval and spatial resolution of the monitored area [12, 13]. For regions larger than $200000\,km^2$, the TWS changes with intervals on month and longer time scale can be monitored, and the accuracy can reach 1.5 cm and above.

The application of GRACE-based TWS changes data is mainly based on the combination of hydrological models and land surface models. GRACE-based TWS retrieval results offer spatial and temporal distribution of vertically integrated water storage (surface water, soil, groundwater, and snowpack) in large river basins. So the errors due to the use of indirect indicators such as the flow rate and precipitation in some hydrological models can be reduced. In initial researches, the GRACE-based TWS is only considered as a reference to the simulation results from hydrological models (such as GDLAS and CPC), which does not involve the uncertainty of hydrological models in the simulation process. Hu et al. [10] proposed that water storage changes in Yangtze River could reach a magnitude of 3.4 cm equivalent water height, with the maximum occurring in the spring and early autumn. Using 5-year GRACE-based TWS change data in China, Zhong et al. [14] pointed out that the water storage in the north-central region of China diminished at an annual rate of 2.4 cm water equivalent height. By combining GRACE and GPS network observation data,

Wang et al. [15] achieved precise retrieval of TWS change in Nordic region and North America. Their studies showed that, over the past decade, a sharp increase appeared in the water storage in North America and Scandinavia and a water recovery process is underway.

With further study, GRACE data is now considered as an important parameter to assess and improve hydrological model simulation. Swenson and Milly [16] and Syed et al. [17] used GRACE data to verify and improve TWS simulation of five climate models and the Global Land Data Assimilation System (GLDAS), respectively. Niu and Yang [18] and Ngo-Duc et al. [19] used GRACE data for the result verification of common land surface model (CLM) of NCAR and ORCHIDEE and corrected the CLM by applying GRACE-based TWS changes data.

Various quantitative analyses between hydrological models and GRACE-based TWS showed that the two simulation results of large areas are in good consistency on a seasonal or longer time scale [16, 20, 21]. Time-series analyses of South and North America, Southeast Asia, the Ganges River Basin, Africa, Eurasia, and other large-scale regions all showed that GRACE-based TWS reflected a significant seasonal variation. However, given the limitations in the spatial resolution of GRACE data, the consistency of hydrological models and GRACE data in smaller areas is not satisfactory [22]. Excluding the errors of GRACE-based TWS change retrieval, the inconsistency is mainly due to the simulation errors of various TWS components from hydrological models. For example, the generalized deviation in estimating the important elements such as groundwater, surface water, and soil water storage by using hydrological models resulted in smaller magnitudes in TWS changes. The errors in the simulation of rainfall, convergence, and other hydrological processes generated phase deviation in TWS changes time series [17, 23, 24]. In addition, by combination of the measured data of rainfall, recharge, soil water, and groundwater, GRACE-based TWS was used for water storage variation of basin on interannual, annual, seasonal, and monthly time scales [25, 26].

Hydrological Components Evaluation

With the improvement of GRACE-based TWS change accuracy, the GRACE data is further used in hydrology and water resources research. For specific areas, TWS changes can be expressed as [10]

$$\frac{dW}{dt} = P - E - R,$$

$$(1)$$

where dW/dt represents TWS changes, and the corresponding equivalent water height can be obtained from GRACE. P represents precipitation, E represents evapotranspiration, and R represents surface runoff. By combining GRACE-based TWS changes, hydrological models, and the measured data, the water storage components including groundwater, soil water, ET, and P – E can be estimated, respectively.

Groundwater estimation is difficult in remote sensing applications in hydrology. Optical remote sensing methods can only combine measured data with spectral data to construct an empirical model for the estimation of groundwater changes, whose groundwater retrieval accuracy is poor. Considering the contribution of groundwater changes to TWS variation, groundwater remote sensing using GRACE was feasible by combination with ancillary measurements of surface waters and soil moisture. Rodell and Famiglietti [27] firstly used GRACE-based TWS change data, soil water, and other auxiliary data to analyze the possibility of monitoring groundwater changes. Swenson et al. [25] studied the relation between measured value of groundwater plus soil water and GRACE-based TWS changes and presented the possibility and sources of error in groundwater estimation based on GRACE data on multiple time scales. Strassberg et al. [28] compared the spatiotemporal correlation between GRACE-based TWS changes and hydrological model simulations and field monitoring data and proposed the uncertainty of GRACE-based groundwater retrieval. The above studies suggest that, for aquifers larger than 450000 km², the accuracy of GRACE-based groundwater can reach 8.7 mm. Rodell et al. applied GRACE data to the groundwater subsidence monitoring in India and found that excessive irrigation and human activities caused the groundwater in the northwest provinces of India to decline at an annual rate of 3.0 cm~4.0 cm [29].

Evapotranspiration (ET) is an important process parameter to studies of hydrology, which is difficult to measure at a regional or continental scale. Although ET can be indirectly estimated using remote sensing data based on empirical models, energy balance models, or physical models (e.g., Penman-Monteith equation), recent remote-sensed ET estimation is far from satisfactory [30]. For ET is a complex process that related to many variables, its estimation uncertainty based on remote sensing data brings errors to ET retrieval results. Based on terrestrial water balance at scale of river basin, changes of regional ET can be estimated by combining TWS change from GRACE with observed precipitation (P) and runoff data (R). Rodell et al. [31] first used the measured precipitation and runoff data to verify the feasibility of GRACE-based ET retrieval. However, the overdependence on measured data restricts the applicability of this method. Therefore, Ramillien et al. [32] improved this algorithm by using the simulated runoff data from hydrological model. The time-series analysis showed that the ET estimation and WGHM simulation were in good consistency. However, uncertainty of GRACE-based ET estimation increases for its overdependence on accuracy of hydrological model simulated variables such as runoff. In addition, Swenson and Wahr [33] combined GRACE-based TWS changes together with the measured runoff data to estimate the difference between precipitation and evapotranspiration (P – E). By comparing with P – E results of a land surface model (GLDAS-Noah), they found that the errors of GLDAS-Noah estimation are mainly due to model force parameter of precipitation.

Drought Analysis

Based on time-series analysis of GRACE-based TWS change with high temporal resolution, extreme hydrological disasters can be monitored and alarmed. Droughts for regional and time-series water storage deficit are the most serious natural hazards that can lead to crop losses and economic havoc in many areas. For droughts can be regarded as terrestrial water storage (TWS) changes that are related to integrated bulk variables, analysis relying upon subcomponents (e.g., precipitation) or proxies (e.g., NDVI and CWSI) of TWS is insufficient. Herewith, another hydrological application of GRACE is severe droughts analysis. Combined with the measurements and hydrological model simulations, Leblanc et al. [34] used GRACE data to detect droughts in Southeastern Australia between 2001 and 2008. Based on time-series analysis of TWS changes in summers from 2002 to 2007, 2005 extreme drought event in the Amazon river basin was detected [35], which was regarded as the worst in over a century. GRACE data is more sensitive to droughts than data-assimilating climate and land surface models such as NECEP and GLDAS, which demonstrated the unique potential of GRACE in monitoring large-scale severe drought events. Validated by two independent hydrological estimates of GLDAS and ECMWF and direct gravity observations from superconducting gravimeters, the 2003 excess terrestrial water storage depletion was observed from GRACE, which can be related to the record-breaking heat wave that occurred in central Europe in 2003 [36]. It indicated that GRACE data can be used in heat wave disaster monitoring and evaluation, whose essence is heat wave caused droughts analysis. Combining with imagery remote sensing methods, Frappart and Wilson et al. applied GRACE data to flood monitoring in several flood zones such as the Mekong River Basin and the Amazon River Basin [37, 38]. However, compared to droughts, the water storage changes based on GRACE data of extreme floods are less responsive, which will bring more uncertainty in GRACE-based floods analysis [39].

Glacier Mass Balance Detection

Driven by global change and population pressure, accelerating of glaciers melting and sea level rise has been a serious global ecological problem. GRACE time-variable gravity field enabled direct measurement of mass loss rates of both mountain glacier and polar glacier systems. As noted by Wouters et al. the GRACE solutions can be used to allow regional estimation of trends, though assessing the mass loss to be dominated by summer events rather than by a linear trend [40]. GRACE-based glacier mass melting volume and rate estimation of Antarctic and Greenland indicate that accelerating polar glacier melting contributes greatly to the current eustatic sea level rise [41–43]. Gardner et al. have also found a rapidly increasing mass loss in the Canadian Arctic Archipelago (CAA) for the period 2004–2009 [44]. Ice loss rate estimation on glaciers of Asia, Alaska, and other global mountains based on GRACE data also agrees with the global tendency of accelerating glacial loss. Accelerated melting of mountain glaciers worldwide might be contributing to the global sea level rise by 0.73 ± 0.10 mm/yr [45, 46]. Moreover, GRACE-based TWS change can be used for researches of different climate variability impact [47, 48] whereas, for the postglacial rebound effects and unique set of gravity data GRACE mission provided, ice sheet mass change estimation includes large uncertainties [49]. Mass change rates estimated by different GRACE solutions may vary by a factor of two or more. So the glacier mass balance detection needs considerable processing to yield usable mass change data [49].

DISCUSSION AND CONCLUSIONS

Through ten years of development, GRACE data has been widely applied in terrestrial hydrology monitoring. Gravity satellite provides new data sets for hydrology researches based on remote sensing technology. Compared with other kinds of hydrological schemes, GRACE can provide realistic spatiotemporal variations of vertically integrated measurement of water storage (groundwater, soil moisture, surface water, snow water, vegetation water, etc.) at the precision of tens of mm of equivalent water height at large scale. However, the hydrological monitoring based on GRACE data needs further research, for example, retrieval accuracy improvement, more quantitative analyses rather than qualitative analyses, and so forth. The focus of future development includes the following aspects.

1. **Improving the Accuracy of Gravity Satellite Measurements**. Because of the limitations of GRACE satellite sensors in orbital altitude, vertical gravity gradient measurements, high frequency signal aliasing, and accurate measurement of time-varying gravity signals, the accuracy and spatial resolution of time-varying Earth gravity field signals at medium

and long wavelength of GRACE is low, which reduced the TWS change retrieval accuracy. Therefore, the development of key technology in gravity satellite sensors to improve accuracy and spatial resolution of satellite monitored gravity field is the basis for a wider application in terrestrial hydrology monitoring.

2. **Water Storage Retrieval Models Research**. Recently, the appropriate spatial resolution for GRACE-based TWS change is 400 km, and the data accuracy is generally 1.5 cm. Limited by recent constraints of GRACE satellite, improvement of retrieval models is the only method for spatial resolution and accuracy retrieval of TWS increases. For example, it is feasible to improve TWS accuracy by the atmosphere and ocean models improvement in data preprocessing, reducing the gravity field changes noises caused by factors unrelated to terrestrial water such as tides and circulation of atmosphere and ocean. In addition, by applying different filtering methods such as anisotropic Gaussian filtering and spherical radial basis function filtering, wavelet analysis in different research areas can also improve the spatial resolution and accuracy of TWS retrieval. Water storage retrieval techniques involve various aspects (e.g., GRACE data preprocessing, gravity field retrieval, and TWS changes estimation) of the conversion of GRACE gravity field data to TWS changes, which are all important in TWS retrieval improvement.

3. **Further Combination of GRACE Data with Associated Hydrological Models**. For GRACE-based TWS changes retrieval does not involve the hydrological mechanisms, it is necessary to complement GRACE-based TWS by hydrological models for furthering its application in terrestrial hydrology monitoring. Further combination of GRACE and hydrological models needs resolving their consistencies of space, time, and component. Spatial consistency can be solved by adjusting the calculation unit of hydrological models to reflect the spatial variability (such as distributed hydrological models). Temporal consistency requires analysis of upscaling GRACE data and downscaling hydrological models simulation. Component consistency can be achieved by GRACE-based TWS signals subdivision or revising parameters of hydrological models to TWS. Through space, time, and component consistency improvement, the gravity field data and measured data together can become basic dataset to force hydrological models in the future.

APPENDICES

A. Water Storage Change Retrieval

The changes in terrestrial water storage result in mass redistribution in the Earth's system, thereby causing changes in the gravity field. For a fixed continental region, the changes in water storage (including soil water and surface snow) come from rainfall, evapotranspiration, river transportation, and deep underground infiltration. Except the rainfall which can cause increased water storage, the remaining three processes all reduce it [10]. Using the FG5 absolute gravimeter, Zhang et al. [50] measured the gravity change of nearly $10^{-7} ms^{-2}$ at the Wuhan University site before and after a rainstorm, which clearly shows the influence of terrestrial water variation on gravity. The Earth's gravity field can be expressed as geoid:

$N(\theta, \lambda)$

$$= a \sum_{n=0}^{\infty} \sum_{m=0}^{n} \left[\overline{C}_{nm} \cos(m\lambda) + \overline{S}_{nm} \sin(m\lambda) \right] \overline{P}_{nm}(\cos\theta),$$
(A.1)

where n and m are harmonic degree and order of the gravity field, respectively; a is the Earth's equatorial radius (about 6,371 km); θ and λ are colatitudes (the difference between 90° and latitude) and longitude; $\overline{C}_{nm}$ and $\overline{S}_{nm}$ are spherical harmonic coefficients (dimensionless); $\overline{P}_{nm}$ is the normalized associated Legendre functions. The maximum value N of the order n of ideal gravity field should be infinite ($N \sim \infty$), while the actual order of spherical harmonic coefficients obtained by the gravity satellite has a finite value ($N < \infty$) and the spatial resolution of gravity field data is estimated approximately to be $\pi a/N$ [51]. Geoid height changes ΔN caused by the movement of substances on the Earth's surface can be expressed as

$\Delta N(\theta, \lambda)$

$$= a \sum_{n=0}^{\infty} \sum_{m=0}^{n} \left[\Delta \overline{C}_{nm} \cos(m\lambda) + \Delta \overline{S}_{nm} \sin(m\lambda) \right] \overline{P}_{nm}(\cos\theta),$$
(A.2)

where $\Delta \overline{C}_{nm}$ and $\Delta \overline{S}_{nm}$ are the changes of n-degree m-order spherical harmonic coefficients of geoid and can be expressed as [5, 52]

$$\left\{\begin{matrix} \Delta\bar{C}_{nm} \\ \Delta\bar{S}_{nm} \end{matrix}\right\}$$

$$= \frac{3}{4\pi a \rho_a (2n+1)}$$

$$\times \int \Delta\rho(r,\theta,\lambda)\,\bar{P}_{nm}(\cos\theta)\left(\frac{r}{a}\right)^{n+2}\left\{\begin{matrix}\cos(m\lambda)\\\sin(m\lambda)\end{matrix}\right\}\sin\theta d\theta\, d\lambda\, dr,$$
$$(A.3)$$

where ρ_a is the average density of the Earth ($5517\,\mathrm{kg/m^3}$) and $\Delta\rho(r,\theta,\lambda)$ is the change of bulk density at a particular location. In gravity field inversion, because the height change of substances on the Earth's surface H is relatively small compared to the Earth's average radius a in gravity field retrieval ((r/a) ≈ 1), the changes of the gravity field directly caused by the surface mass can be expressed as

$$\left\{\begin{matrix} \Delta\bar{C}_{nm} \\ \Delta\bar{S}_{nm} \end{matrix}\right\}_{\mathrm{surf}}$$

$$= \frac{3}{4\pi a \rho_a (2n+1)}$$

$$\times \int \Delta\sigma(\theta,\lambda)\,\bar{P}_{nm}(\cos\theta)\left\{\begin{matrix}\cos(m\lambda)\\\sin(m\lambda)\end{matrix}\right\}\sin\theta d\theta\, d\lambda,$$
$$(A.4)$$

where $\Delta\sigma$ is surface density change $\Delta\sigma = \int \Delta\rho(r,\theta,\lambda)dr$. Furthermore, the surface mass variation in loads will cause deformation to the solid Earth, which in turn can indirectly cause variations to the gravity field. This can be expressed as [5, 52]

$$\left\{\begin{matrix} \Delta\bar{C}_{nm} \\ \Delta\bar{S}_{nm} \end{matrix}\right\}_{\mathrm{soild}}$$

$$= \frac{3k'_n}{4\pi a \rho_a (2n+1)}$$

$$\times \int \Delta\sigma(\theta,\lambda)\,\bar{P}_{nm}(\cos\theta)\left\{\begin{matrix}\cos(m\lambda)\\\sin(m\lambda)\end{matrix}\right\}\sin\theta d\theta\, d\lambda,$$
$$(A.5)$$

where k'_n is load LOVE number of coefficients, and the specific value of k'_n can be found in relevant literature [11]. Thus, the change of the Earth's gravity field caused by mass variations on the Earth's surface is given by

$$
\left\{ \begin{array}{c} \Delta \overline{C}_{nm} \\ \Delta \overline{S}_{nm} \end{array} \right\} = \left\{ \begin{array}{c} \Delta \overline{C}_{nm} \\ \Delta \overline{S}_{nm} \end{array} \right\}_{surf} + \left\{ \begin{array}{c} \Delta \overline{C}_{nm} \\ \Delta \overline{S}_{nm} \end{array} \right\}_{soild}
$$

$$
= \left(1 + k'_n \right) \left\{ \begin{array}{c} \Delta \overline{C}_{nm} \\ \Delta \overline{S}_{nm} \end{array} \right\}_{surf} .
$$
(A.6)

If the spherical harmonic expansion is performed on the surface density change $\Delta \sigma$, then

$$
\Delta \sigma \left(\theta, \lambda \right)
$$

$$
= a\rho_w \sum_{n=0}^{\infty} \sum_{m=0}^{n} \left[\widehat{C}_{nm} \cos \left(m\lambda \right) + \widehat{S}_{nm} \sin \left(m\lambda \right) \right] \overline{P}_{nm} \left(\cos \theta \right),
$$
(A.7)

where ρw is the density of water and can be considered as mass variation on the Earth's surface expressed in water equivalent height. Wahr et al. [5] proposed that

$$
\left\{ \begin{array}{c} \Delta \widehat{C}_{nm} \\ \Delta \widehat{S}_{nm} \end{array} \right\} = \frac{\rho_a}{3\rho_w} \frac{2n+1}{1+k'_n} \left\{ \begin{array}{c} \Delta \overline{C}_{nm} \\ \Delta \overline{S}_{nm} \end{array} \right\} .
$$
(A.8)

The surface density change $\Delta \sigma$ can be calculated using the following equation:

$$
\Delta \sigma \left(\theta, \lambda \right)
$$

$$
= \frac{a\rho_a}{3} \sum_{n=0}^{\infty} \sum_{m=0}^{n} \frac{2n+1}{1+k'_n} \left[\overline{C}_{nm} \cos \left(m\lambda \right) + \overline{S}_{nm} \sin \left(m\lambda \right) \right]
$$

$$
\times \overline{P}_{nm} \left(\cos \theta \right).
$$
(A.9)

For ocean and atmosphere mass variations are removed based on Parallel Ocean Program (POP) model, the above equation is the basic equation for the retrieval of surface mass variations based on spatiotemporal changes gravity field. The Earth's surface density changes can be derived from the changes of gravity field coefficients obtained from GRACE satellites.

B. Evaluation Errors

Currently, the latest available GRACE dataset product is the RL05. The accuracy of data provided by CSR and GFZ is better than that provided by JPL. The evaluation errors of GRACE are brought from satellite instruments measurement, retrieval models, and other factors as follows.

1. Gravity field data measured by GRACE satellite may be contaminated by satellite measurement errors for the influence of satellite orbit, satellite K-band ranging, and accelerometer measurement [20]. The satellite measurement errors also include poor accuracy of C20 due to the insensitivity of track geometry to the gravity field's low degree gravitational variations [5, 20]. It is generally released with removing variation of C20 [10] or replacing it by satellite laser ranging (SLR) substitution [20]. In addition, the missing of first-degree spherical harmonic coefficients will also bring error to GRACE. It can commonly be resolved by substituting value calculated from the term of the seasonal changes of the Earth's mass center [53, 54] or ignoring its impact [55].

2. In theory, the retrieval of gravity field variation needs to use spherical harmonic coefficients of all degrees from 0 to infinity. However, gravity satellites can only provide definite order data. So the surface density change $\Delta\sigma$ in retrieval models is treated by spherical harmonic expansion to definite orders. For the impact of high-order terms on the Earth's surface density change $\Delta\sigma$ cannot be ignored, it results in truncation errors in gravity field retrieval. RL05 water storage data is estimated by CSR using a retrieval model truncated to degree 60 [8]. Zhu et al. (2008) compared the global water storage retrieval results using models truncated to 15 degrees, 20 degrees, 35 degrees, and 60 degrees [56]. They found that although some information of TWS change may be missing the retrieval result of TWS change became more marked with lower truncation degree. Generally, the water storage retrieval truncated to order 60 is widely adopted [17].

3. For terrestrial water monitoring focusing on the mass changes of a particular area (e.g., river basin), it requires the integral process on density change $\Delta\bar{\sigma} = ((\int \Delta\sigma(\theta, \lambda)u(\theta, \lambda)d\Omega)/\Omega)$. The function of regional characteristics (θ, λ) is equal to 1 inside the particular area and to zero outside. The error will be brought for the discontinuity of (θ, λ) in the domain of integration. In addition, for the influence of rapid increase of the errors of GRACE gravity field model coefficients with increase of spherical harmonic coefficients degree, signal leakage errors, and the striping pollution [23], filtering methods are proposed to smooth GRACE data for noise reducing, which will result in filtering errors of

retrieval results. Proposed filtering methods include spatial averaging, symmetric Gaussian filtering, optimized decorrelation filtering, time-series method, global hydrological model correction method, kernel-independent component analysis, and optimal smoothing kernel method [53, 57–62]. Using filtering for the integral treatment of surface density changes can effectively remove the striping to a certain extent. However, the obtained average surface density is critical for it reduces the useful energy of geophysical signals and results in filtering errors. In addition, the existing filtering methods require the support of a priori knowledge (such as filtering radius and truncation degree). Therefore, in the actual retrieval process, the filter selection and parameter calibration require an understanding of the specific regional characteristics [60, 63].

4. TWS changes retrieval based on spatiotemporal variations of gravity field remains considerable uncertainty, which includes errors produced in the removal of tidal movement and the mass migration because of the atmosphere and ocean circulation [11], as well as the errors in hydrological models used for the estimation of other terrestrial water parameters [6].

ACKNOWLEDGMENTS

This project was supported by National Natural Science Foundation of China (Grant no. 51309210) and the Chinese Academy of Sciences (Grant no. KZZD-EW-08).

REFERENCES

1. Y. Huang, D. Jiang, D. Zhuang, Y. Zhu, and J. Fu, "An improved approach for modeling spatial distribution of water use profit—a case study in Tuhai Majia Basin, China," Ecological Indicators, vol. 36, pp. 94–99, 2014.

2. Y.-H. Huang, D. Jiang, D.-F. Zhuang, J.-H. Wang, H.-J. Yang, and H.-Y. Ren, "Evaluation of relative water use efficiency (RWUE) at a regional scale: a case study of Tuhai-Majia Basin, China," Water Science and Technology, vol. 66, no. 5, pp. 927–933, 2012.

3. J. S. Famiglietti, "Remote sensing of terrestrial water storage, soil moisture and surface waters," in The State of the Planet: Frontiers and Challenges in Geophysics, R. S. J. Sparks and C. J. Hawkesworth, Eds., pp. 197–207, 2004.

4. B. D. Tapley, S. Bettadpur, M. Watkins, and C. Reigber, "The gravity recovery and climate experiment: mission overview and early results,"

Geophysical Research Letters, vol. 31, no. 9, Article ID L09607, 2004.

5.   J. Wahr, S. Swenson, V. Zlotnicki, and I. Velicogna, "Time-variable gravity from GRACE: first results,"Geophysical Research Letters, vol. 31, no. 11, Article ID L11501, 2004.

6.   A. Güntner, "Improvement of global hydrological models using GRACE data," Surveys in Geophysics, vol. 29, no. 4-5, pp. 375–397, 2008.

7.   G. Ramillien, J. S. Famiglietti, and J. Wahr, "Detection of continental hydrology and glaciology Signals from GRACE: a review," Surveys in Geophysics, vol. 29, no. 4-5, pp. 361–374, 2008.

8.   D. P. Chambers, "Converting 11 Release-04 Gravity Coefficients into Maps of Equivalent Water Thickness," 2007, http://gracetellus.jpl.nasa.gov/files/GRACE-dpc200711_RL04.pdf.

9.   M. Becker, W. LLovel, A. Cazenave, A. Güntner, and J.-F. Crétaux, "Recent hydrological behavior of the East African great lakes region inferred from GRACE, satellite altimetry and rainfall observations,"Comptes Rendus Geoscience, vol. 342, no. 3, pp. 223–233, 2010.

10.   X. Hu, J. Chen, Y. Zhou, C. Huang, and X. Liao, "Seasonal water storage change of the Yangtze River basin detected by Grace," Science in China D, vol. 49, no. 5, pp. 483–491, 2006.

11.   J. Wahr, M. Molenaar, and F. Bryan, "Time variability of the Earth's gravity field: hydrological and oceanic effects and their possible detection using GRACE," Journal of Geophysical Research B, vol. 103, no. 12, pp. 30205–30229, 1998.

12.   M. Rodell and J. S. Famiglietti, "Detectability of variations in continental water storage from satellite observations of the time dependent gravity field," Water Resources Research, vol. 35, no. 9, pp. 2705–2723, 1999.

13.   M. Rodell and J. S. Famiglietti, "An analysis of terrestrial water storage variations in Illinois with implications for the Gravity Recovery and Climate Experiment (GRACE)," Water Resources Research, vol. 37, no. 5, pp. 1327–1339, 2001.

14.   M. Zhong, J. Duan, H. Xu, P. Peng, H. Yan, and Y. Zhu, "Trend of China land water storage redistribution at medi- and large-spatial scales in recent five years by satellite gravity observations,"Chinese Science Bulletin, vol. 54, no. 5, pp. 816–821, 2009.

15.   H. Wang, L. Jia, H. Steffen et al., "Increased water storage in North America and Scandinavia from GRACE gravity data," Nature Geoscience, vol. 6, no. 1, pp. 38–42, 2013.

16.   S. C. Swenson and P. C. D. Milly, "Climate model biases in seasonally of continental water storage revealed by satellite gravimetry," Water Resources Research, vol. 42, no. 3, Article ID W03201, 2006.

17.   T. H. Syed, J. S. Famiglietti, M. Rodell, J. Chen, and C. R. Wilson, "Analysis of terrestrial water storage changes from GRACE and GLDAS," Water Resources Research, vol. 44, no. 2, Article ID W02433, 2008.

18.   G.-Y. Niu and Z.-L. Yang, "Assessing a land surface model's improvements with GRACE estimates,"Geophysical Research Letters, vol. 33, no. 7, Article ID L07401, 2006.

19.   T. Ngo-Duc, K. Laval, G. Ramillien, J. Polcher, and A. Cazenave, "Validation of the land water storage simulated by Organising Carbon and Hydrology in Dynamic Ecosystems (ORCHIDEE) with Gravity Recovery and Climate Experiment (GRACE) data," Water Resources Research, vol. 43, no. 4, Article ID W04427, 2007.

20.   J. L. Chen, C. R. Wilson, B. D. Tapley, and J. C. Ries, "Low degree gravitational changes from GRACE: validation and interpretation," Geophysical Research Letters, vol. 31, no. 22, Article ID L22607, pp. 1–5, 2004.

21.   R. Klees, E. A. Revtova, B. C. Gunter et al., "The design of an optimal filter for monthly GRACE gravity models," Geophysical Journal International, vol. 175, no. 2, pp. 417–432, 2008.

22.   K.-W. Seo, C. R. Wilson, J. S. Famiglietti, J. L. Chen, and M. Rodell, "Terrestrial water mass load changes from Gravity Recovery and Climate Experiment (GRACE)," Water Resources Research, vol. 42, no. 5, Article ID W05417, 2006.

23.   B. D. Tapley, S. Bettadpur, J. C. Ries, P. F. Thompson, and M. M. Watkins, "GRACE measurements of mass variability in the Earth system," Science, vol. 305, no. 5683, pp. 503–505, 2004.

24.   D. P. Lettenmaier and J. S. Famiglietti, "Hydrology: water from on high," Nature, vol. 444, no. 7119, pp. 562–563, 2006.

25.   S. Swenson, P. J.-F. Yeh, J. Wahr, and J. Famiglietti, "A comparison of terrestrial water storage variations from GRACE with in situ measurements from Illinois," Geophysical Research Letters, vol. 33, no. 16, Article ID L16401, 2006.

26.   L. Xavier, M. Becker, A. Cazenave, L. Longuevergne, W. Llovel, and O. C. R. Filho, "Interannual variability in water storage over 2003–2008 in the Amazon Basin from GRACE space gravimetry, in situ river level and

precipitation data," Remote Sensing of Environment, vol. 114, no. 8, pp. 1629–1637, 2010.

27. M. Rodell and J. S. Famiglietti, "The potential for satellite-based monitoring of groundwater storage changes using GRACE: the High Plains aquifer, Central US," Journal of Hydrology, vol. 263, no. 1–4, pp. 245–256, 2002.

28. G. Strassberg, B. R. Scanlon, and M. Rodell, "Comparison of seasonal terrestrial water storage variations from GRACE with groundwater-level measurements from the High Plains Aquifer (USA)," Geophysical Research Letters, vol. 34, no. 14, Article ID L14402, 2007.

29. M. Rodell, I. Velicogna, and J. S. Famiglietti, "Satellite-based estimates of groundwater depletion in India," Nature, vol. 460, no. 7258, pp. 999–1002, 2009.

30. X. Lu and Q. Zhuang, "Evaluating evapotranspiration and water-use efficiency of terrestrial ecosystems in the conterminous United States using MODIS and AmeriFlux data," Remote Sensing of Environment, vol. 114, no. 9, pp. 1924–1939, 2010.

31. M. Rodell, J. S. Famiglietti, J. Chen et al., "Basin scale estimates of evapotranspiration using GRACE and other observations," Geophysical Research Letters, vol. 31, no. 20, Article ID L20504, 2004.

32. G. Ramillien, F. Frappart, A. Güntner, T. Ngo-Duc, A. Cazenave, and K. Laval, "Time variations of the regional evapotranspiration rate from Gravity Recovery and Climate Experiment (GRACE) satellite gravimetry," Water Resources Research, vol. 42, no. 10, Article ID W10403, 2006.

33. S. Swenson and J. Wahr, "Estimating large-scale precipitation minus evapotranspiration from GRACE satellite gravity measurements," Journal of Hydrometeorology, vol. 7, no. 2, pp. 252–270, 2006.

34. M. Leblanc, P. Tregoning, G. Ramillien, S. O. Tweed, and A. Fakes, "Basin-scale, integrated observations of the early 21st century multiyear drought in southeast Australia," Water Resources Research, vol. 45, no. 4, Article ID W04408, 2009.

35. J. L. Chen, C. R. Wilson, B. D. Tapley, Z. L. Yang, and G. Y. Niu, "2005 drought event in the Amazon River basin as measured by GRACE and estimated by climate models," Journal of Geophysical Research B, vol. 114, no. 5, Article ID B05404, 2009.

36. O. B. Andersen, S. I. Seneviratne, J. Hinderer, and P. Viterbo, "GRACE-derived terrestrial water storage depletion associated with the 2003

European heat wave," Geophysical Research Letters, vol. 32, no. 18, Article ID L18405, pp. 1–4, 2005.

37. F. Frappart, K. Do Minh, J. L'Hermitte et al., "Water volume change in the lower Mekong from satellite altimetry and imagery data," Geophysical Journal International, vol. 167, no. 2, pp. 570–584, 2006.

38. M. D. Wilson, P. Bates, D. Alsdorf et al., "Modeling large-scale inundation of Amazonian seasonally flooded wetlands," Geophysical Research Letters, vol. 34, no. 15, Article ID L15404, 2007.

39. S.-C. Han, C. K. Shum, C. Jekeli, and D. Alsdorf, "Improved estimation of terrestrial water storage changes from GRACE," Geophysical Research Letters, vol. 32, no. 7, Article ID L07302, pp. 1–5, 2005.

40. B. Wouters, D. Chambers, and E. J. O. Schrama, "GRACE observes small-scale mass loss in Greenland,"Geophysical Research Letters, vol. 35, no. 20, Article ID L20501, 2008.

41. I. Velicogna and J. Wahr, "Measurements of time-variable gravity show mass loss in Antarctica,"Science, vol. 311, no. 5768, pp. 1754–1756, 2006.

42. D. C. Slobbe, P. Ditmar, and R. C. Lindenbergh, "Estimating the rates of mass change, ice volume change and snow volume change in Greenland from ICESat and GRACE data," Geophysical Journal International, vol. 176, no. 1, pp. 95–106, 2009.

43. P. L. Svendsen, O. B. Andersen, and A. A. Nielsen, "Acceleration of the Greenland ice sheet mass loss as observed by GRACE: confidence and sensitivity," Earth and Planetary Science Letters, vol. 364, pp. 24–29, 2013.

44. A. S. Gardner, G. Moholdt, B. Wouters et al., "Sharply increased mass loss from glaciers and ice caps in the Canadian Arctic Archipelago," Nature, vol. 473, no. 7347, pp. 357–360, 2011.

45. J. L. Chen, B. D. Tapley, and C. R. Wilson, "Alaskan mountain glacial melting observed by satellite gravimetry," Earth and Planetary Science Letters, vol. 248, no. 1-2, pp. 353–363, 2006.

46. K. Matsuo and K. Heki, "Time-variable ice loss in Asian high mountains from satellite gravimetry,"Earth and Planetary Science Letters, vol. 290, no. 1-2, pp. 30–36, 2010.

47. D. García-García, C. C. Ummenhofer, and V. Zlotnicki, "Australian water mass variations from GRACE data linked to Indo-Pacific climate variability," Remote Sensing of Environment, vol. 115, no. 9, pp. 2175–2183, 2011.

48. W. W. Immerzeel, L. P. H. van Beek, and M. F. P. Bierkens, "Climate change will affect the asian water towers," Science, vol. 328, no. 5984, pp. 1382–1385, 2010.

49. L. S. Sørensen, S. B. Simonsen, K. Nielsen et al., "Mass balance of the Greenland ice sheet (2003–2008) from ICESat data—the impact of interpolation, sampling and firn density," Cryosphere, vol. 5, no. 1, pp. 173–186, 2011.

50. W. Zhang, Y. Wang, and C. Zhang, "The preliminary analysis of effects of the soil moisture on gravity observations," Cartographica Sinica, vol. 30, no. 2, pp. 108–111, 2001.

51. G. Ramillien, J. S. Famiglietti, and J. Wahr, "Detection of continental hydrology and glaciology signals from GRACE: a review," Surveys in Geophysics, vol. 29, no. 4-5, pp. 361–374, 2008.

52. J. L. Chen, C. R. Wilson, R. J. Eanes, and B. D. Tapley, "Geophysical contributions to satellite nodal residual variation," Journal of Geophysical Research B, vol. 104, no. 10, pp. 23237–23244, 1999.

53. J. K. Willis, D. P. Chambers, and R. S. Nerem, "Assessing the globally averaged sea level budget on seasonal to interannual timescales," Journal of Geophysical Research C, vol. 113, no. 6, Article ID C06015, 2008.

54. E. W. Leuliette and L. Miller, "Closing the sea level rise budget with altimetry, Argo, and Grace,"Geophysical Research Letters, vol. 36, no. 4, Article ID L04608, 2009.

55. J. L. Chen, C. R. Wilson, J. S. Famiglietti, and M. Rodell, "Attenuation effect on seasonal basin-scale water storage changes from GRACE time-variable gravity," Journal of Geodesy, vol. 81, no. 4, pp. 237–245, 2007.

56. G. Zhu, J. Li, H. Wen, and J. Wang, "Study on variations of global continental water storage with GRACE gravity field models," Journal of Geodesy and Geodynamics, vol. 28, no. 5, pp. 39–44, 2008.

57. S. Swenson and J. Wahr, "Methods for inferring regional surface-mass anomalies from Gravity Recovery and Climate Experiment (GRACE) measurements of time-variable gravity," Journal of Geophysical Research B, vol. 107, no. 9, pp. 3-1–3-13, 2002.

58. S. Swenson and J. Wahr, "Monitoring changes in continental water storage with grace," Space Science Reviews, vol. 108, no. 1-2, pp. 345–354, 2003.

59. K.-W. Seo and C. R. Wilson, "Simulated estimation of hydrological loads from GRACE," Journal of Geodesy, vol. 78, no. 7-8, pp. 442–456, 2005.

60. S. Swenson and J. Wahr, "Post-processing removal of correlated errors in GRACE data," Geophysical Research Letters, vol. 33, no. 8, Article ID L08402, 2006.

61. G. Ramillien, F. Frappart, A. Cazenave, and A. Güntner, "Time variations of land water storage from an inversion of 2 years of GRACE geoids," Earth and Planetary Science Letters, vol. 235, no. 1-2, pp. 283–301, 2005.

62. F. Frédéric, G. Ramillien, M. Leblanc, et al., "An independent component analysis filtering approach for estimating continental hydrology in the GRACE gravity data," Remote Sensing of Environment, vol. 115, no. 1, pp. 187–204, 2011.

63. S. Werth, A. Güntner, R. Schmidt, and J. Kusche, "Evaluation of GRACE filter tools from a hydrological perspective," Geophysical Journal International, vol. 179, no. 3, pp. 1499–1515, 2009.

Chapter 5

CLIMATE CHANGE DETECTION AND MODELING IN HYDROLOGY

Saeid Eslamian[1], Kristin L. Gilroy[2] and Richard H. McCuen[2]

[1] Isfahan University of Technology, Isfahan, Iran
[2] University of Maryland, USA

INTRODUCTION

Detection of a change is defined as the process of demonstrating that climate or a system affected by climate has changed in some defined statistical sense, without providing a reason for that change. Attribution is defined as the process of evaluation of the relative contribution of multiple causal factors to a change or event with an assignment of statistical confidence. However, the observed changes must be able to be detected (IPCC 2010).

Attribution to a change in climatic conditions includes the assessments that attribute an observed change in a variable of interest to a specific observed change in climate conditions based on the process knowledge and relative importance of a change in climate condition in determining the observed impacts (Hao et al. 2008; Liu and Xia 2011). The associated confidence levels should be evaluated for the data, model, methods, and the factors used in the study (IPCC 2010).

Seibert et al. (2010) used the three different approaches for change detection modeling employing a modified version of the HBV (Hydrologiska Byråns Vattenbalansavdelning) model (Bergstrom 1976,1992) to conclude that catchment-scale runoff increases following severe wildfire. The application of the HBV model as a change detection tool indicated the increases in peak flows following severe wildfire and the related road building and harvesting of the dead and damaged forest vegetations.

The parameter uncertainty of various parameter sets is commonly known in hydrologic and climatologic modeling. It is an issue seldom addressed in modeling approaches for detecting changes (Pappenberger and Beven 2006;

Seibert and McDonnell 2010). Employing a large number of parameter sets rather than a single set of parameter values facilitates the assessment of the associated uncertainty.

The detection of climate change impacts on the observed climate and elements of the hydrological cycle have made a great progress, recently (Amiri and Eslamian, 2010). Based on the climate model simulation, the optimal methods have been used to detect the responses of observed change to Green House Gas emissions from the other external forcing at large spatial scales. Presently, the detection of anthropogenic influence is not yet possible for all of the climate variables. It is still difficult to attribute the observed changes in climate or variables of interest on a spatial scale lower than five thousands kilometers and temporal scales of less than fifty years. For the basin aquifers recharged by precipitation or surplus irrigation and influenced by artificial and strong human activities, the detection studies mainly focused on analytical approaches to link physical impacts to changes in temperature or precipitation as a tool. For the basins with better observational data and more sensitivity towards climate change, the use of formal detection methods to identify the pattern responses of the hydrological cycle to external forcing is a valuable and promising area of further research (Liu and Xia 2011).

Xoplaki et al. (2008) investigated data requirement for climate change detection and modeling research in the Mediterranean. They indicated that data availability allows the validation of scientific results on climate change detection and attribution.

The need for long measurements of climate and hydrologic data for studies of climate variability and change is very important. New et al. (1999, 2000) have developed the fields for many climate variables and it is essential to develop these further and extend them to some hydrological variables such as discharge and runoff, for both climate variability and change studies and also climate model validation.

Most of the investigations in climate variability and change detection have focused on only temperature, due to well representation by the available network. The temperature measurement exhibits the relatively high correlation decay lengths. Both upcoming impacts and those of previous events are, however, much more dependent upon the changes in precipitation. The changes are not only vital for hydrology, but are also much more important than temperature for many other sectors, such as agriculture and range management. The studies of large-scale changes in precipitation are hampered by the requirement to obtain access to considerably more precipitation data than is conventionally available. A similar case can also be met for runoff data. Climate change detection studies need to be undertaken on a global scale, and both the available networks of runoff and precipitation data are inadequate. Presently, the best that can be achieved are the investigations on the regional and catchment scales (Cihlar et al. 2000).

A method for detecting the impacts of disturbance on catchment-scale hydrology is the paired catchment approach. The method combines rainfall-runoff modeling to account for natural fluctuations in daily streamflow, uncertainty analyses using the generalized likelihood uncertainty estimation method to identify and separate hydrologic model uncertainty from unexplained variation, and GLS regression change detection models to provide a formal experimental framework for detecting changes in daily streamflow relative to variations in daily hydrologic and climatic data (Zégre et al. 2010).

Precipitation data indicate both increasing and decreasing trends for different regions of the world.Zhang et al. (2007) detected the human influence on twentieth century precipitation trends.

The main objective of this study is to describe the statistical techniques for detecting changes in hydrological events. The flood records are selected for this purpose. Statistical tests and distributions, significance levels and confidence intervals, risk and uncertainty and nonstationarity are discussed in detail for the flood series.

DETECTION OF CHANGE IN FLOOD RECORDS

Graphical analysis is generally the first attempt at detecting change in a flood record. Unfortunately, the natural variation of year-to-year flooding greatly exceeds the variation expected due to climate change of recent years. Thus, the latter would likely not be visually evident from a graphical portrayal of a flood record. In comparison, it takes a considerable level of urban development before the hydrologic effects of urbanization can be graphically detected, especially if the trend is temporally gradual rather than abrupt. Whether change is due to global warming or urbanization, we can not be certain whether the nonstationarity factor will cause a change in the probability distribution of floods or just its moments. Thus, more sophisticated methods of detection are needed.

Commonly, the next step in detection of change is with statistical methods. Some of the problems with statistical detection of the effects of climate change include uncertainty in the distribution from which the sample was drawn, outliers, poorly measured values, no knowledge as to when the climate change began to significantly influence flooding, and the compounding effects of land cover change such as deforestation. In addition to these factors, identifying a statistically significant change requires the specification of a statistical level of significance. The value selected is a central factor in statistical decision making, yet a systematic way of identifying the optimum level of significance is not known. The selection of a level of significance is not a trivial decision as the power of the test will depend on the level selected.

Eslamian et al. (2009) investigated to detect an existing trend in wind speed and to evaluate the effect of climate change on frequency analysis of wind speed in Iran. The purpose of this study was to present the recent trends and variations in measured wind speed at twenty-two gauging stations along the whole country of Iran. In addition, the effect of climate change was evaluated in frequency analysis and heterogeneity. For understanding wind behavior in time periods, the trend test and frequency analysis were performed for evaluating wind magnitude and duration.

SELECTION OF STATISTICAL METHOD TO DETECT TREND

Statistical methods are generally designed to be most sensitive to one type of change, such as the change in central tendency, the change in dispersion, or the change in the statistical distribution. Change can be gradual or abrupt and different statistical methods should be applied to such data. A change in a data set that is characterized by gradually varying flows may not be detected if the statistical method applied is more sensitive to abrupt change. Evidence does not currently exist as to the distributional effect that climate change introduces to a flood record. For example, will prolonged climate change cause annual maximum floods to follow a Generalized Extreme Value (GEV) distribution rather than the commonly accepted log-Pearson type III distribution (LP3)? Bulletin 17B, which was developed to estimate flood frequencies and, therefore, flood risk, assumes that hydrologic data follows a LP3 (Interagency 1982). However, many recent studies in regards to precipitation data are based on other distributions. For example, Koutsoyiannis (2004), Stedinger (2000), Gellens (2002), and Karin and Zwiers (2005) selected the GEV distribution to model extreme events whileWilby and Wigley (2002) and Semenov and Bengtsson (2002) chose the gamma distribution to model daily precipitation events. Therefore, if agreement does not currently exist on the appropriate distribution to represent hydrologic data, it will be difficult to determine the appropriate distribution as the concept of nonstationarity is introduced. To compound the problem, climate change is expected to be gradual and thus, the distribution may be subject to continual change.

In addition to the appropriate distribution, the effects of climate change on the moments of a distribution are unknown. The studies have suggested that climate change will increase the more intense rainfalls but have little effect on total annual rainfalls (Hennessy et al. 1997; Karl and Knight 1997; Wilby and Wigley 2002). Kharin and Zwiers (2005) found a significant change in the location and scale parameters for the GEV distribution in a global analysis of precipitation extremes; however, the effects of climate change are expected to vary regionally. Therefore, global analyses may not be applicable at the regional level. Since the expected changes in the statistical distribution as well as the moments are not known and, therefore, must be assumed, this reduces the ability of statistical tests to effectively decide whether or not change has occurred. Without this knowledge, the best statistical method to detect climate change can not be selected without recognizing the importance of this type of uncertainty.

THE ASSUMPTION OF A START TIME

Before the problem of modeling can be solved, the first issue that must be addressed is detection of change. The most obvious question is: when did the effect of climate change begin to significantly influence the hydrologic variable of interest, e.g., annual maximum peaks? For example, Olsen et al. (1999) varied the start and end dates of flood records analyzed for gauges in the Missouri and Mississippi River basin. Based on the linear regression results, they found that different record lengths within the same flood record influenced the significance of the trend detected. Therefore, knowledge of the time at which nonstationarity began is necessary in order to correctly identify trends.

Identifying the start time of nonstationarity is important because the time that climate change is assumed to have become influential will influence the model used to represent the hydrologic change. If incorrectly selected, the model type can greatly affect projected changes in hydrologic data. For example, let's assume that we know that climate change will introduce a linear trend in the hydrologic variable. If the start date is quite uncertain, then the slope of the linear trend will be biased depending on the assumed start time. An incorrectly assumed early start time will lead to an underpredicted slope. Likewise, assuming a late start time would result in a relatively steep slope and long-term overprediction. Figure 1 shows the mean annual discharge (cfs) for the USGS gauge 05464500 at Cedar Rapids, Iowa. A linear trend was fit to the data with two start times: 1903 and 1960, represented by the solid and dashed regression lines, respectively. If extrapolated to the year 2050, the model based on a 1960 start time projects a mean discharge that is 12% greater than the model based on a later start time. Therefore, the uncertainty of predictions due to inaccurate start times can be significant. A statistical test that is sensitive to the start date needs to have high statistical power. Otherwise, incorrect start times will result, with the subsequent impacts on models, future projections, and risk estimation.

The Anacostia River at Hyattsville, MD, can be used to illustrate the effect of start time on the trend of peak discharge rates. Figure 2 shows the annual maximum peak discharge from 1939 to 1988. In the late 1950's, urban development influenced flood flows, with the effect apparent in Figure 2. Linear models were used to model the trend in flood peaks as a function of time:

$$1955\text{-}1988: q_p = 344 + 54.11 * t \qquad t=1 \text{ in } 1955 \tag{1}$$

$$1960\text{-}1988: q_p = 4178 + 29.57 * t \qquad t=1 \text{ in } 1960 \tag{2}$$

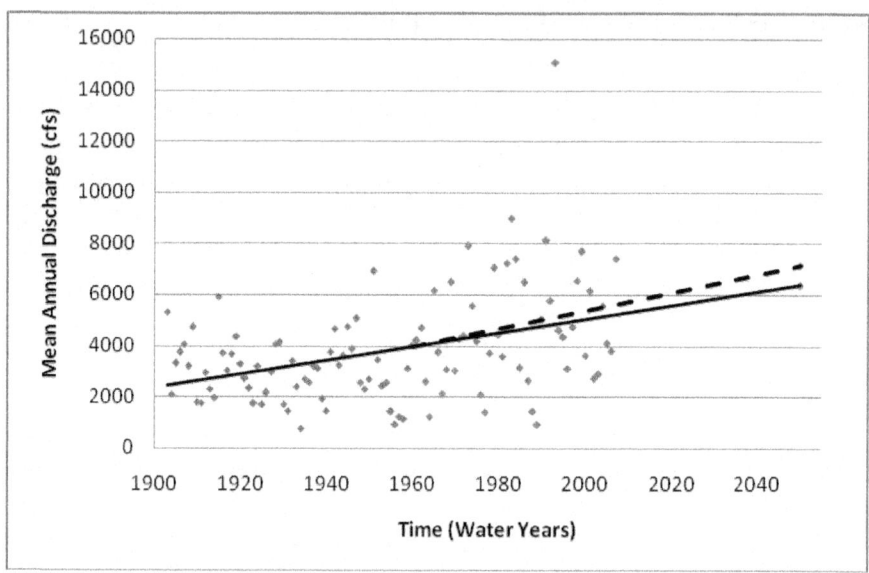

Figure 1. Variation in Trend Modeled based on Different Start Times for the Mean Annual Discharge (cfs) for the USGS Cedar Rapids Gauge (05464500) with the Dashed and Solid Line Representing a 1960 and 1903 Start Time, Respectively

While the record lengths are similar (34 and 29 years, respectively), the equations are quite different. The estimated floods for the year 2011 would be 6528 cfs (185 m³/s) and 5716 cfs (162 m³/s) for Eqs. 1 and 2, respectively. This represents a difference of 13% based solely on the start time. This example illustrates the sensitivity of estimated flood magnitudes to the start time for modeling time trends.

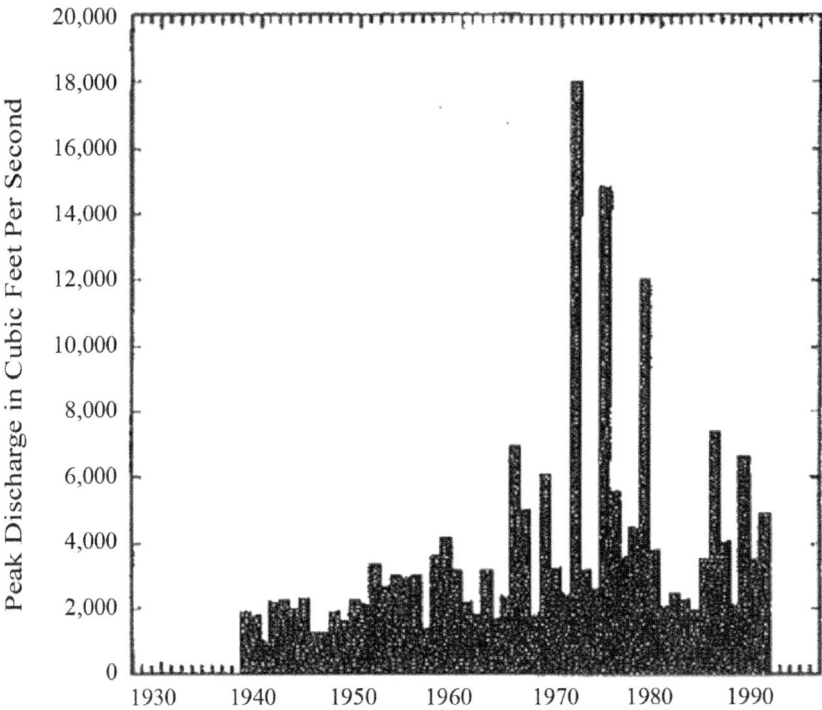

Figure 2. Peak Discharge Data for Annacostia River at Hyattsville, Maryland.

Assuming that the start time can be reasonably estimated, the next question of interest is: What affect will climate change have on the physical processes that determine the nature or characteristics of the climate change? Will the climatic change influence the statistics of the hydrologic variable, e.g., increase the mean or the variance, or will it change the distribution, e.g., from a LP3 to a GEV? The accuracy of projected discharges will greatly depend on the change assumed, which will subsequently influence the accuracy of risk estimates.

SELECTION OF STATISTICAL DISTRIBUTIONS

As mentioned in the discussion of statistical method selection, the appropriate distribution for hydrologic data is unknown. While this leads to difficulties in trend detection, it also influences the projection of hydrologic events, such as flooding. Bulletin 17B currently recommends the LP3 distribution; however, the GEV distribution is recommended by many studies as well (Martins and Stedinger 2000). Both distributions represent extreme data; however, the extreme events projected by each distribution can vary. For example, Figure 3 compares the frequency curve fit to the Cedar Rapids annual maximum peak discharge at the USGS gauge 05464500 with both the LP3 and GEV distributions. The GEV distribution projects a 100-yr flood that is 12.6% greater than that of the LP3 distribution. Therefore, depending on the distribution selected, engineers may over or underestimate the 100-yr storm designing flood management structures. Therefore, it is important that the correct probability distributed is selected for hydrologic data for design and policy development.

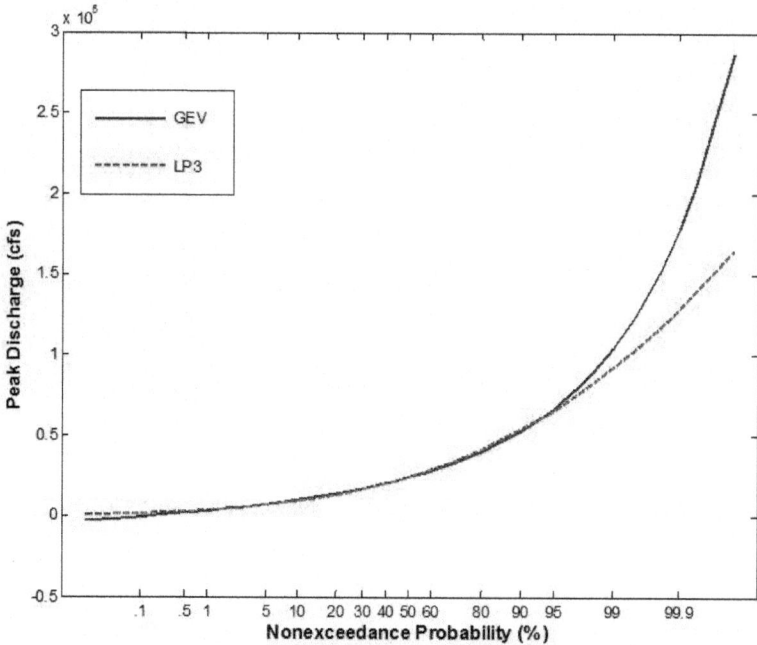

Figure 3. Frequency Curve for LP3 and GEV Distribution of Cedar Rapids Annual Maximum Peak Discharge Data.

SELECTION OF STATISTICAL MODELS

In addition to the start time and distribution selection, uncertainties in future predictions will result from the model form selected to represent the change being analyzed. For example, evidence points to a nonlinear trend in hydrologic data as the result of climate change, but some global models suggest an increasing function while other models suggest a decreasing function because of policies that control CO_2 emissions. Likewise, even when urbanization is known to influence measured flood magnitudes, it has been difficult to identify the model structure that approximates the temporal effects of changes in the physical processes associated with urban land cover change. Linear trends are often assumed as other more complex functions do not lead to greater accuracy. Yet, the assumed model structure will dictate the magnitude of floods projected for future land cover conditions. Assuming an incorrect function form to represent the trend of increasing flood discharge rates will influence peak discharge estimated for the future. This is another source of uncertainty as an incorrect model structure can lead to overprediction or underprediction of design floods and their associated risks.

To illustrate the potential effect of model structure on floods estimated for future times, the flood series for the Anacostia River (see Figure 2) was fitted for the 1955-1988 period using a linear model (Eq. 1) and the following power or log-linear model:

$$1955\text{-}1988: \quad q_p = 2384 * t^{0.153} \tag{3}$$

Based on this power model form, the 2011 estimated discharge would be 4425 cfs (125 m³/s), which differs from the discharge estimated using Eq. 1 by 2103 cfs (59.5 m³/s), or 38.4%. The effect of model structure is significant and this issue is must be considered in an attempt to model the effects of climate change on hydrologic data.

This same problem will influence the accuracy of modeling the effects of hydrologic nonstationarity due to a changing climate. It is difficult to even detect whether or not climate change has introduced systematic variation into a flood record let alone identifying the structural form of the temporal change induced by the climate change. Much effort will need to be expended on the detection of change and to identify the best model structure will be a central modeling issue. The model structure finally adopted will greatly influence assessments of future flood risk and the design of hydrologic and hydraulic infrastructure with design lives that will cover the period of climate change.

CONFIDENCE INTERVALS UNDER CHANGING CONDITIONS

The third issue important to the modeler and to policy makers is: How can confidence intervals be computed on projected discharges when the distribution and parameters of future discharges are unknown? Given the lack of certainty in the distribution of climate-affected discharges, the most obvious choice of methods for computing confidence intervals would be those used for linear regression analysis. This approach requires a minimum of inputs, such as the standard error of estimate, the sample size of the existing record, and the standard deviation of the time variable. One problem with this approach is the lack of stationarity. Confidence intervals computed using traditional methods assume stationarity. A new approach will be needed. Uncertainty associated with the nonstationarity will likely lead to much wider confidence intervals on hydrologic variables such as peak discharge rates. An approach based on Monte Carlo simulation for different levels of nonstationarity may be necessary to produce more accurate assessments of the confidence of projected discharges.

With the current state of the art, nonparametric methods are the generally accepted approach to detection of change. Numerous tests are available, but many of these lack statistical power. For example, the Runs Test, which was designed to assess the presence or lack of randomness, i.e., independence, could be applied over portions of a flood record to identify the portions of the record that were not homogeneous. If the other causative factors, such as urbanization can be ruled out, then the test may detect an approximate time at which nonstationarity began. Given the low statistical power of the test, the best estimate of the start date will likely be very imprecise.

The Kendall Tau Test is one of the more commonly used tests for detecting nonrandomness. This test is often preferred because it is designed for data with a monotonically increasing trend, as opposed to an episodic change. It can be applied to either long or short flood records, although the accuracy of the decision will depend on the record length.

Tests for serial independence, such as the Pearson Test and the nonparametric Spearman Test, can be effective for identifying the existence of trends. However, when the Spearman Test is applied to hydrologic data where the data are ordered by year of occurrence, the critical values generally available do not apply. Some analyses have correlated the hydrologic variable with the integer of time, i.e., 1 to n, used as the second variable. The Spearman Test statistic has a different distribution function when the integer of time is applied as one of the variables rather than the adjacent value of the discharge value (Conley and McCuen 1997). Both of these tests should use only the peak discharge sequence rather than correlating discharge and time.

LEVEL OF SIGNIFICANCE FOR DETECTION DECISIONS

The above are all important issues to those involved in assessing the effects of climate change, yet they may not be the most important issue. Regardless of the distribution assumed or the statistical test selected, the significance of an effect will depend on the level of significance adopted for decision making. Karl and Knight (1997) used the 5% level of significance to determine whether increases in precipitation within the United States were significant in the 20th century. Burns and Elnur (2002)reported hydrologic trends detected based on a 10% level of significance. Olsen et al. (1999) identified trends detected in flood records with both a 1% and 5% levels of significance. Traditionally, a

5% level is used, but evidence that this is really appropriate for hydrologic variables has not been addressed. It is unlikely that a 5% level of significance would lead to detection of hydrologic change due to a changing climate, as the sampling variation is generally quite dominant and would overwhelm the effect of climate change. Additionally, use of a 5% level will likely lead to a test having low statistical power. For example, Figures 3a and b display two time series simulated based on normally distributed errors and the same intercept and slope coefficients; however, the standard error for was increased by 100% from the Figure 3a to the Figure 3b time series. The Kendall Tau Test was applied to each data set and Z-values equal to 2.85 and 1.31 were calculated for the data in Figure 3a and 3b, respectively. Therefore, at the 5% level of significance, the null hypothesis was rejected for Figure 3aand accepted for Figure 3b. The null hypothesis was rejected at the 10% level of significance forFigure 3a. Therefore, the variation within the data influences the level of significance at which a trend will be detected, which is a concern when dealing with variables that contain high variation, such as hydrologic data. Before any statistical test is adopted, the issue of statistical power and the level of significance needs to be studied.

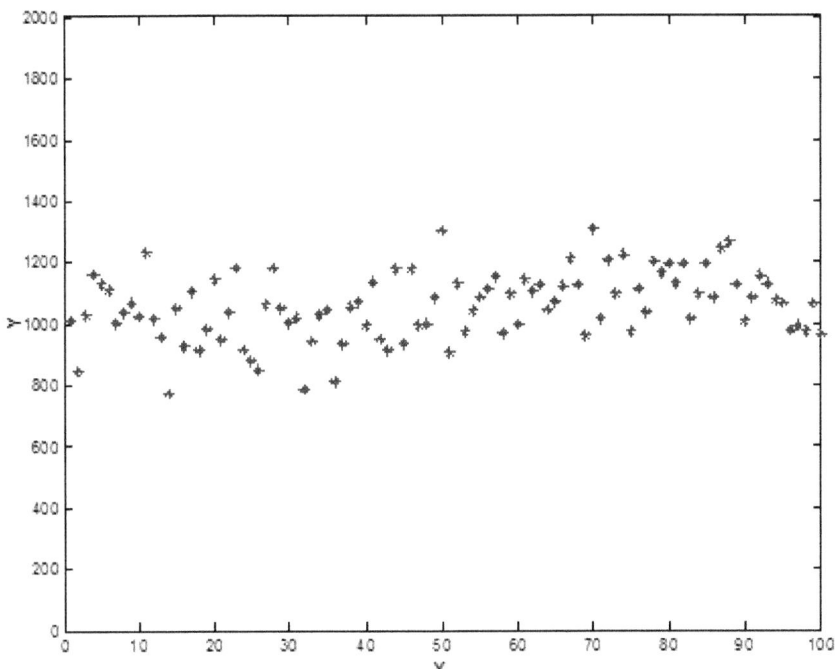

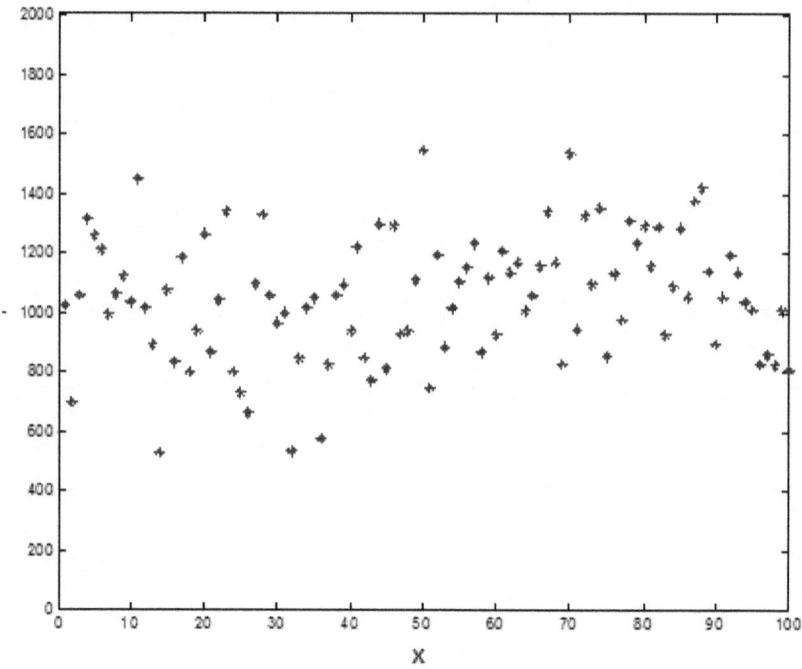

Figure 3A,B. a and b. Simulated Time Series with Intercept = 1000, Slope = 1.25, and Se = 100 and 200., respectively, with Resulting Kendall Tau Statistics equal to 1.31 and 2.85, respectively.

The null hypothesis of interest to this issue is: climate change has not introduced nonstationarity into the flood series. The alternative hypothesis would be that the flood series is nonstationary. Using a small level of significance of 5% or 1% gives some assurance that a true null hypothesis will not be falsely rejected, but it also increases the chance of not identifying an effect and accepting the null hypothesis, when, if fact, it is false. A 5% level of significance will then likely lead to not making an adjustment of the flood series for nonstationarity, whereas use of a higher level of significance would dictate making such an adjustment. It seems that adjusting a series with a minimal effect of climate change would be preferable to failing to make a needed adjustment, even if the adjustment is small. Thus, the proper level of significance to be used in climate change analyses need to be investigated.

NONSTATIONARITY AND FLOOD RISK

Engineering designs are commonly based on an estimate of the 100-yr discharge, where the discharge is based on an analysis of the historic flood record or on a regression model fitted using regional flood records. Assuming that global climate change models are correct and that extreme rainfalls are expected to increase over time, then it is reasonable to assume that the time series of annual maximum discharges will increase over the next century. Under these conditions, storms of the size on which a design was based will occur more frequently. Thus, the 100-yr discharge of future times will be larger than the current 100-yr discharge. Likewise, under nonstationary conditions the current 100-yr flood event will occur more frequently (Olsen et al. 1998). The likelihood of the bridge opening passing runoff magnitudes will decrease, which means that the risk of failure will continuously increase with time.

Engineering design and risk assessments need to consider this nonstationarity of the T-yr discharge, where T is the return period used in design, e.g., T = 100yrs. A design made in 2010 based on the 100-yr discharge assessed using current information and knowledge will not have the same risk of failure as the climate changes. This should be considered in the design. If the design life for the 2010 project is 50 years, it may not be appropriate to design for the estimated 100-yr event for 2010 meteorological and hydrological conditions as then the project would be underdesigned. Similarly, designing for 2060 climate conditions would assume overdesign for each of the 49 years between 2010 and 2059. The optimal design discharge under these nonstationary global climate conditions would need to account for the rate of change of discharge over time. As many climate change scenarios show an increasing trend with time, the nonlinearity of the nonstationarity would require a temporally adjusted risk analysis.

Analyses have shown that the location and scale parameters of annual maximum flood series are expected to increase with increasing global climate change. These would raise the frequency curve and increase the exceedence probability of a flood magnitude. This is easily shown using a binomial risk analysis. Consider the case of a site where the 2010 conditions indicate a flood skew of 0.3 with the log moments shown in Table 1 for the decades that define the design life of the project. Assume that a project is designed to control the 100-yr flood magnitude of 1411 cms (49821 cfs). Assuming climate change will cause the log moments to increase as shown, then the return period of the design discharge increases over the 30 year period from the current 100 years to a 41-year event in 30 years. The binomial risk for each decade would change from 9.56% in the first decade, to 14% in the second decade and 19.1% in the third decade. Therefore, over the design life of the project, the project as designed has an increased likelihood of being exceeded. This change in the expected exceedence probability would provide a benefit-cost ratio of the project that was less than the ration on which a design based on stationary conditions would provide. Failure to account for the effect of climate change in the design would lead to long-term under design.

This conclusion is not intended to suggest that the project should be designed to the 100-yr flood condition at the end of the design life, as this would reflect a long-term design that would exceed the 100-yr protection. For example, if the project were designed for the 2040 flood moments, with a discharge of 2733 cms (96521 cfs), then the annual exceedence probability for current conditions would suggest a design exceedence probability of 0.00383, which reflects a return period of 261 years. Except in the last year of the design life of the project, the facility would have a protection that exceeds the required value. Using this value in a project benefit-cost ratio would provide a value that would exceed the long-term ratio that could be expected over the design life.

Table 1. Binomial Risk over Time

Decade	Log mean	Log sd	Flow (cfs)	Flow (cms)	K	p	T (yrs)	Prob.
2010	3.12	0.62	49821	1411	2.544	0.0100	100	0.0956
2020	3.15	0.63	56605	1603	2.456	0.0126	79	0.1074
2030	3.21	0.65	73070	2068	2.288	0.0177	57	0.1416
2040	3.28	0.67	96521	2733	2.116	0.0242	41	0.1908

If the intent is to provide, on average, 100-yr protection, then an integrated procedure is needed. Such a method would need to consider the temporally changing flood potential at the site, as indicated by the changing moments. The continually changing flood risk would need to be estimated. A method developed to integrate the effect of the changing flood risk would be expected to account for these changes in flood potential.

CONCLUSIONS AND RECOMMENDATIONS

An important benefit of the modeling approach is that, in addition to quantification of change resulting from a disturbance, comparison of model parameters between pre- and post-event periods provides an indication of hydrological processes alteration by a severe event.

Detecting the effect of climate change in measured hydrologic data is a difficult, but important, task (Eslamian 2006). It has ramifications to assessing flood risk, the design of water resource infrastructure, and the avoidance of assigning too much weight to other nonstationary factors, such as urbanization, that contribute to hydrologic change. If the effect of climate change is even marginally significant, but not accounted for when attempting to assess the effects of urbanization in hydrologic data, then the effects of urbanization will likely be overstated. The results of such analyses will lead to biased designs. Infrastructure design that fails to account for climate change can be inadequate to meet the safety needs of a community, as the likelihood of severe flooding will increase because of climate change. Thus, floods that occur over the design life will likely be larger and more frequent than designed for. These situations are central to the issue of assessing flood risk, which has obvious implications to public safety, resource allocation, and the disruption of facility use.

The implications of global warming are significant, as policies will be made to address the issue and climate change may have significant economic effects. Therefore, uncertainties in projections of the effects of climate change must be considered in designing infrastructure, establishing public policies, and in economic decisions. A few of the uncertainties have been discussed in this paper, with an emphasis on uncertainties related to climate change modeling. Projecting to the future represents extrapolation, and given the uncertainties in modeling procedures, data, and our theoretical knowledge of the underlying processes, decision makers must consider these uncertainties. Record lengths of data used to calibrate climate models are short and contain very significant levels of nonsystematic variation. Such uncertainty will be carried over to projections made to the year 2100. The nonsystematic variation often reflects our lack of a full understanding of causal factors. Efforts through research need to be made to reduce these uncertainties in order to increase the accuracy of projections into the future.

Given the uncertainties of and interactions between the different model parameters, such explanations need to be approached with caution. Nevertheless, these suggestions of altered processes can direct further investigation and hypothesis formulation.

The following observations are especially important for the further climate variability investigations:

- Daily runoff series for a few hundred smaller natural catchments, about 1000 km² catchment area, employed for research purpose distributed over the globe.
- Monthly runoff series for the top few hundred catchments around the world, possibly having natural flows.
- Long time series of hydrological records

Impacts of climate change on water quality are also largely determined by hydrological changes and by the nature of pollutants as flushing or dilution-controlled. The most significant impact of urban development on water resources is an increase in overall surface runoff and the flashiness of the associated storm hydrograph. The increase in impervious surface area associated with urban development also contributes to degradation of water quality as a result of non-point source pollution. The modelling studies on the combined impacts of climate change and urban development have found that either change may be more significant, depending on scenario assumptions and basin characteristics, and that each type of change may amplify or ameliorate the effects of the other (Praskievicz and Chang 2009).

REFERENCES

1. M. J. Amiri, S. S. Eslamian, 2010 Investigation of climate change in Iran, Journal of Environmental Science and Technology, 34208216

2. S. Bergstrom, 1976 Development and application of a conceptual runoff model for Scandinavian catchments". SMHI Reports RHO, 7Norrko ping, Sweden, 134.

3. S. Bergstrom, 1992 The HBV model-its structure and applications". SMHI Reports RH, 4Norrko ping, Sweden.

4. D. H. Burn, M. A. H. Elnur, 2002 Detection of Hydrologic Trends and Variability". Journal of Hydrology, 255107122

5. J. Cihlar, W. Grabs, J. Landwehr, 2000 Establishment of a Global Hydrological Observation Network for Climate, Report of the GCOS/ GTOS/HWRP Expert Meeting, Geisenheim, Germany, 2630June. Report GCOS 63, Report GTOS 26, Secretariat of the World Meteorological Organization, WMO /TD- 1047

6. L. C. Conley, R. H. Mc Cuen, 1997 Modified Critical Values for Spearman's Test of Serial Correlation". Journal of Hydrologic Engineering, 23133135

7. S. S. Eslamian, H. Hasanzadeh, 2009 Detecting and Evaluating Climate Change Effect on Frequency Analysis of Wind Speed in Iran, International Journal of Global Energy Issues, Special Issue on Wind Modelling and Frequency Analysis (WMFA). 323

8. S. S. Eslamian, 2006 Detection of Hydrologic Changes, International Symposium on Drylands Ecology and Human Security, Dubai, United Arab Emirates.

9. X. Hao, Y. Chen, C. Xu, et al.2008 Impacts of climate change and human activities on the surface runoff in the Tarim River Basin over the last fifty years". Water Resources Management, 2211591171

10. K. J. Hennessy, J. M. Gregory, J. F. B. Mitchell, 1997 Changes in Daily Precipitation under Enhanced Greenhouse Conditions". Climate Dynamics, 13667680

11. Interagency Advisory Committee on Water Data1982 Guidelines for Determining Flood Flow Frequency", Bulletin 17B of the Hydrology Committee, USGS, Office of Water Data Coordination, Reston, VA.

12. IPCC2010 Meeting Report of the Intergornmental Panel on Climate Change Expert Meeting on Detection and Attribution Related to Anthropogenic Climate Change". IPCC Working Group I Technical Support Unit, University of Bern, Switzerland, 55pp.

13. T. R. Karl, R. W. . Knight, 1998 Secular Trends of Precipitation Amount, Frequency, and Intensity in the United States". Bulletin of the American Meteorological Society, 792231241

14. V. V. Kharin, F. W. . Zwiers, 2005 Estimating Extremes in Transient Climate Change Simulations". Journal of Climate, 1811561173

15. D. Koutsoyiannis, 2004 Statistics of Extremes and Estimation of Extreme Rainfall: II. Empirical Investigation of Long Rainfall records". Hyrdological Sciences Journal, 494591610

16. C. Liu, J. Xia, 2011 Detection and Attribution of Observed Changes in the Hydrological Cycle under Global Warming". Advances in Climate Change Research, 213137

17. E. S. Martins, J. R. Stedinger, 2000 Generalized Maximum-Likelihood Generalized Extreme-Value Quantile Estimators for Hydrologic Data". Water Resources Research, 363737744

18. M. New, M. Hulme, P. D. Jones, 1999 Representing twentieth-century space-time climate variability, Part I: Development of a 1961-90 mean monthly terrestrial climatology". Journal of Climate, 12829856

19. M. New, M. Hulme, P. D. Jones, 2000 Representing twentieth-century space-time climate variability, Part II: Development of 1901-1996 monthly grids of terrestrial surface climate". Journal of Climate, 1322172238

20. J. R. Olsen, J. H. Lambert, Y. Y. Haimes, 1998 Risk of Extreme Events Under Nonstationary Conditions". Risk Analysis, 184497510

21. J. R. Olsen, J. R. Stedinger, N. C. Matalas, E. Z. Stakhiv, 1999 Climate Variability and Flood Frequency Estimation for the Upper Mississippi and Lower Missouri Rivers". Journal of the American Water Resources Association, 35615091524

22. F. Pappenberger, K. J. Beven, 2006 Ignorance is bliss: or seven reasons not to use uncertainty analysis". Water Resources Research, 42(5): W05302.

23. S. Praskievicz, H. Chang, 2009 A review of hydrological modelling of basin- scale climate change and urban development impacts". Progress in Physical Geography, 335650671

24. J. Seibert, J. J. Mc Donnell, R. D. Woodsmith, 2010 Effects of wildfire on catchment runoff response: A modeling approach to change detection". Hydrology Research, 415378390

25. J. Seibert, J. J. Mc Donnell, 2010 Land-cover impacts on streamflow: A change detection modeling approach that incorporates parameter uncertainty", Hydrological Sciences Journal, 553316332

26. V. A. Semenov, L. Bengtsson, 2002 Secular Trends in Daily Precipitation Characteristics: Greenhouse Gas Simulation with a Coupled AOGCM". Climate Dynamics, 19123140

27. R. L. Wilby, T. M. L. Wigley, 2002 Future changes in the distribution of daily precipitation totals across North America." Geophys. Res. Lett., 29: 1135, doi:10.1029/2001GL013048.

28. E. Xoplaki, A. Toreti, F. G. Kuglitsch, J. Luterbacher, 2008 Data availability in the Mediterranean: requirements for climate change detection and modeling research, June, MEDARE, Proceedings of the International Wokshop on Rescue and Digitization of Climate Records in the Mediterranean Basin, University of Bern, Switzerland.

29. N. Zégre, A. E. Skaugset, N. A. Som, J. J. Mc Donnell, L. M. Ganio, 2010 In lieu of the paired catchment approach: Hydrologic model change detection at the catchment scale". Water Resources Research, 46: W11544 EOFPP., doi:10.1029/2009WR008601.

30. X. Zhang, F. W. Zwiers, G. C. Hegeri, et al.2007 Detection of human influence on twentieth century precipitation trends. Nature, 448461465

Chapter 6

RECURRENT NEURAL NETWORK BASED APPROACH FOR SOLVING GROUNDWATER HYDROLOGY PROBLEMS

Ivan N. da Silva[1], José Ângelo Cagnon[2] and Nilton José Saggioro[3]

[1] University of São Paulo (USP), São Carlos, SP, Brazil

[2] São Paulo State University (UNESP), Bauru, SP, Brazil

[3] University of São Paulo (USP), Bauru, SP, Brazil

INTRODUCTION

Many communities obtain their drinking water from underground sources called aquifers. Official water suppliers or public incorporations drill wells into soil and rock aquifers looking for groundwater contained there in order to supply the population with drinking water. An aquifer can be defined as a geologic formation that will supply water to a well in enough quantities to make possible the production of water from this formation. The conventional estimation of the exploration flow involves many efforts to understand the relationship between the structural and physical parameters. These parameters depend on several factors, such as soil properties and hydrologic and geologic aspects [1].

The transportation of water to the reservoirs is usually done through submerse electrical motor pumps, being the electric power one of the main sources to the water production. Considering the increasing difficulty to obtain new electrical power sources, there is then the need to reduce both operational costs and global energy consumption. Thus, it is important to adopt appropriate operational actions to manage efficiently the use of electrical power in these groundwater hydrology problems. For this purpose, it is essential to determine a parameter that expresses the energetic behavior of whole water extraction set, which is here defined as Global Energetic Efficiency Indicator (GEEI). A methodology using artificial neural networks is here developed in order to take into account several experimental tests related to energy consumption in submerse motor pumps.

The GEEI of a depth is given in Wh/m3.m. From a dimensional analysis, we can observe that the smaller numeric value of GEEI indicates the better energetic efficiency to the water extraction system from aquifers.

For such scope, this chapter is organized as follows. In Section 2, a brief summary about water exploration processes are presented. In Section 3, some aspects related to mathematical models applied to water exploration process are described. In Section 4 is formulated the expressions for defining the *GEEI*. The neural approach used to determine the *GEEI* is introduced in Section 5, while the procedures for estimation of aquifer dynamic behavior using neural networks are presented in Section 6. Finally, in Section 7, the key issues raised in the chapter are summarized and conclusions are drawn.

WATER EXPLORATION PROCESS

An aquifer is a saturated geologic unit with enough permeability to transmit economical quantities of water to wells [10]. The aquifers are usually shaped by unconsolidated sands and crushed rocks. The sedimentary rocks, such as arenite and limestone, and those volcanic and fractured crystalline rocks can also be classified as aquifers.

After the drilling process of groundwater wells, the test known as *Step Drawdown Test* is carried out. This test consists of measuring the aquifer depth in relation to continue withdrawal of water and with crescent flow on the time. This depth relationship is defined as *Dynamic Level* of the aquifer and the aquifer level at the initial instant, i.e., that instant when the pump is turned on, is defined as *Static Level*. This test gives the maximum water flow that can be pumped from the aquifer taking into account its respective dynamic level. Another characteristic given by this test is the determination of *Drawdown Discharge Curves*, which represent the dynamic level in relation to exploration flow [2]. These curves are usually expressed by a mathematical function and their results have presented low precision.

Since aquifer behavior changes in relation to operation time, the *Drawdown Discharge Curves* can represent the aquifer dynamics only in that particular moment. These changes occur by many factors, such as the following: i) aquifer recharge capability; ii) interference of neighboring wells or changes in its exploration conditions; iii) modification of the static level when the pump is turned on; iv) operation cycle of pump; and v) rest time available to the well. Thus, the mapping of these groundwater hydrology problems by conventional identification techniques has become very difficult when all above considerations are taken into account. Besides the aquifer behavior, other components of the exploration system interfere on the global energetic efficiency of the system.

On the other hand, the motor-pump set mounted inside the well, submersed on the water that comes from the aquifer, receives the whole electric power supplied to the system. From an eduction piping, which also supports physically the motor pump, the water is transported to the ground surface and from there, through an adduction piping, it is transported to the reservoir, which is normally located at an upper position in relation to the well. To transport water in this hydraulic system, it is necessary several accessories (valves, pipes, curves, etc.) for its implementation. Figure 1 shows the typical components involved with a water extraction system by means of deep wells.

The resistance to the water flow, due to the state of the pipe walls, is continuous along all the tubing, and will be taken as uniform in every place where the diameter of the pipe to be constant.

This resistance makes the motor pump to supply an additional pressure (or a load) in order to water can reach the reservoir. Thus, the effect created by this resistance is also called "load loss along the pipe". Similar to the tubing, other elements of the system cause a resistance to the fluid flow, and therefore, load losses. These losses can be considered local, located, accidental or singular, due to the fact that they come from particular points or parts of the tubing.

Regarding the hydraulic circuit, it is observed that the load loss (distributed and located) is an important parameter, and that it varies with the type and the state of the material.

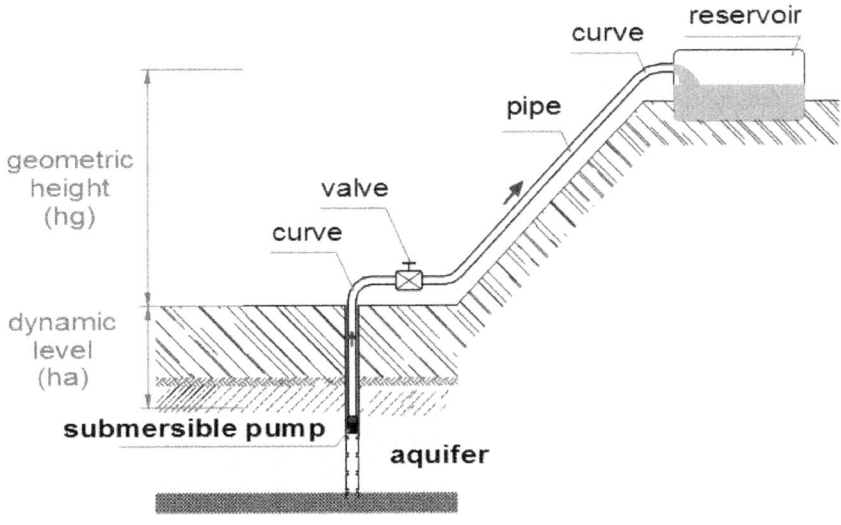

Figure 1. Components of the pumping system.

Therefore, old tubing, with aggregated incrustation along the operational time, shows a load loss different of that present in new tubing. A valve turned off twice introduces a bigger load loss than that when it is totally open. A variation on the extraction flow also creates changes on the load loss. These are some observations, among several other points, that could be done.

Another important factor concerning the global energetic efficiency of the system is the geometrical difference of level. However, this parameter does not show any variation after the total implantation of the system. Concerning this, two statements can be done: i) when mathematical models were used to study the lowering of the piezometric surface, these models should frequently be evaluated in certain periods of time; ii) the exploration flow of the aquifer assumes a fundamental role in the study of the hydraulic circuit and it should be carefully analyzed.

In order to overcome these problems, this work considers the use of parameters, which are easily obtained in practice, to represent the capitation system, and the use of artificial neural networks to determine the exploration flow. From these parameters, it is possible to determine the *GEEI* of the system.

MATHEMATICAL MODELS APPLIED TO WATER EXPLORATION PROCESS

One of the most used mathematical models to simulate aquifer dynamic behavior is the Theis' model [1,9]. This model is very simple and it is used to transitory flow. In this model, the following hypotheses are considered: i) the aquifer is confined by impermeable formations, ii) the aquifer structure is homogeneous and isotropic in relation to its hydro-geological parameters, iii) the aquifer thickness is considered constant with infinite horizontal extent, and iv) the wells penetrate the entire aquifer and their pumping rates are also considered constant in relation to time.

The model proposed by Theis can be represented by the following equations:

$$\frac{\partial^2 s}{\partial r^2} + \frac{1}{r} \cdot \frac{\partial s}{\partial r} = \frac{S}{T} \cdot \frac{\partial s}{\partial t}$$

(1)

$$s(r,0) = 0$$

(2)

$$s(\infty,t) = 0$$

(3)

$$\lim_{r \to 0}[r(\frac{\partial s}{\partial r})] = -\frac{Q}{2 \cdot \pi \cdot T} \tag{4}$$

where:

s is the aquifer drawdown;

Q is the exploration flow;

T is the transmissivity coefficient;

r is the horizontal distance between the well and the observation place.

Applying the Laplace's transform on these equations, we have:

$$\frac{d^2 s^-}{dr^2} + \frac{1}{r} \cdot \frac{ds^-}{dr} = \frac{S}{T} \cdot w \cdot s^- \tag{5}$$

$$s^-(r,w) = A \cdot K_0 \cdot (r \cdot \sqrt{(S/T)} \cdot w) \tag{6}$$

$$\lim_{r \to 0}[r(\frac{ds}{dr})] = -\frac{Q}{2 \cdot \pi \cdot T \cdot w} \tag{7}$$

where:

w is the Laplace's parameter;

S is the storage coefficient.

Thus, the aquifer drawdown in the Laplace's space is given by:

$$s^-(r,w) = \frac{q}{2 \cdot \ \cdot T} \cdot \frac{K_0 \cdot (r \cdot \sqrt{(S/T) \cdot w})}{w} \tag{8}$$

This equation in the real space is as follows:

$$h - h_0(r,t) = s(r,t) = \frac{Q}{2 \cdot \ \cdot T} \cdot L^{-1} \left[\frac{K_0 \cdot (r \cdot \sqrt{(S/T) \cdot w})}{w} \right] \tag{9}$$

The Theis' solution is then defined by:

$$S = \frac{q}{4 \cdot \ \cdot T} \int_u^\infty \frac{e^{-y}}{y} dy = \frac{Q}{4 \cdot \ \cdot T} W(u) \tag{10}$$

where:

$$u = \frac{r^2 \cdot S}{4 \cdot T \cdot t} \tag{11}$$

Finally, from Equation (10), we have:

$$W(u) = 2 \cdot L^{-1} \left[\frac{K_0 \cdot (r \cdot \sqrt{(S/T) \cdot w})}{w} \right],$$

(12)

where:

L^{-1} is the Laplace's inverse operator.

K_0 is the hydraulic conductivity.

From analysis of the Theis' model, it is observed that to model a particular aquifer is indispensable a high technical knowledge on this aquifer, which is mapped under some hypotheses, such as confined aquifer, homogeneous, isotropic, constant thickness, etc. Moreover, other aquifer parameters (transmissivity coefficient, storage coefficient and hydraulic conductivity) to be explored must be also defined. Thus, the mathematical models require expert knowledge of concepts and tools of hydrogeology.

It is also indispensable to consider that the aquifer of a specific region shows continuous changes in its exploration conditions. The changes are normally motivated by the companies that operate the exploration systems, by drilling of new wells or changes of the exploration conditions, or still, motivated by drilling of illegal wells. These changes have certainly required immediate adjustment on the Theis' model. Another fact is that the aquifer dynamic level modifies in relation to exploration flow, operation time, static level, and obviously with those intrinsic characteristics of the aquifer under exploration. In addition, neighboring wells will also be able to cause interference on the aquifer.

Therefore, although to be possible the estimation of aquifer behavior using mathematical models, such as those presented in [11]-[16], they present low precision because it is more difficult to consider all parameters related to the aquifer dynamics. For these situations, intelligent approaches [17]-[20] have also been used to obtain a good performance.

DEFINING THE GLOBAL ENERGETIC EFFICIENCY INDI-CATOR

As presented in [3], "Energetic Efficiency" is a generalized concept that refers to set of actions to be done, or then, the description of reached results, which become possible the reduction of demand by electrical energy. The energetic efficiency indicators are established through relationships and variables that can be used in order to monitor the variations and deviations on the energetic efficiency of the systems. The descriptive indicators are those that characterize the energetic situation without looking for a justification for its variations or deviations.

The theoretical concept for the proposed Global Energetic Efficiency Indicator will be presented using classical equations that show the relationship between the absorbed power from the electric system and the other parameters involved with the process.

As presented in [3], the power of a motor-pump set is given by:

$$P_{mp} = \frac{\gamma \cdot Q \cdot H_T}{75 \cdot \eta_{mp}}$$
(13)

where:

P_{mp} is the power of the motor-pump set (CV);

$\gamma\gamma$ is the specific weight of the water (1000 kgf/m^3);

Q is the water flow (m^3/s);

HT is the total manometric height (m);

$\eta_{mp}\eta_{mp}$ is the efficiency of the motor-pump set ($\eta_{motor} \cdot \eta_{pump}\eta_{motor} \cdot \eta_{pump}$).

Substituting the following values {1 CV $\cong\cong 736$ Watts; 1 m^3/s = 1/3600 m^3/h; $\gamma\gamma = 1000$ kgf/m^3 } inequation (13), we have:

$$P_{mp} = \frac{2.726 \cdot Q \cdot H_T}{\eta_{mp}}$$
(14)

The total manometric height (HT) in elevator sets to water extraction from underground aquifers is given by:

$$H_T = H_a + H_g + \Delta h\, f_t$$
(15)

where:

HT is the total manometric height (m);

Ha is the dynamic level of the aquifer in the well (m);

Hg is the geometric difference in level between the well surface and the reservoir (m);

$\Delta h f t$ is total load loss in the hydraulic circuit (m).

From analyses on the variables in (15), it is observed that only the variable corresponding to the geometric difference in level (Hg) can be considered constant, while other two will change along the operation time of the well.

The dynamic level (Ha) will change (to lower) since the beginning of the pumping until the moment of stabilization. This observation is verified in short period of time, as for instance, a month. Besides this variation, which can present a cyclic behavior, it is possible that other types of variation, due to interferences from other neighboring wells, can take place as well as alterations in the aquifer characteristics.

The total load loss will also vary during the pumping, and it is dependent on hydraulic circuit characteristics (diameter, piping length, hydraulic accessories, curves, valves, etc.).

These characteristics can be considered constant, since they usually do not change after installed. However, the total load loss is also dependent on other characteristic of the hydraulic circuit, which frequently changes along the useful life of the well. These variable characteristics are given by: i) roughness of the piping system, ii) water flow, and iii) operational problems, such as semi-closed valves, leakage, etc.

Observing again Figure 1, it is verified that the necessary energy to transport the water from the aquifer to the reservoir, overcoming all the inherent load losses, it is supplied by the electric system to the motor-pump set. Thus, using these considerations and substituting (15) in (14), we have:

$$P_{el} = \frac{2.726 \cdot Q \cdot (H_a + H_g + \Delta h \, f_t)}{\eta_{mp}} \tag{16}$$

where:

P_{el} is the electric power absorbed from electric system (W);

Q is the water flow (m³/h);

H_a is the dynamic level of the aquifer in the well (m);

H_g is the geometric difference of level between the well surface and the reservoir (m);

Δh_{ft} is the total load loss in the hydraulic circuit (m);

η_{mp} is the efficiency of the motor-pump set ($\eta_{motor} \cdot \eta_{pump}$).

From (16) and considering that an energetic efficiency indicator should be a generic descriptive indicator, the *Global Energetic Efficiency Indicator* (*GEEI*) is here proposed by the following equation:

$$GEEI = \frac{P_{el}}{Q \cdot (H_a + H_g + \Delta h \, f_t)} \tag{17}$$

Observing equation (17), it is verified that the *GEEI* will depend on electric power, water flow, dynamic level, geometric difference of level, and total load loss of the hydraulic circuit.

The efficiency of the motor-pump set does not take part in (17) because its behavior will be reflected inversely by the *GEEI*. Thus, when the efficiency of the motor-pump set is high, the *GEEI* will be low. Therefore, the best *GEEI* will be those presenting the smallest numeric values.

Another reason to exclude the efficiency of the motor-pump set in (17) is the difficulty to obtain this value in practice. Since it is a fictitious value, it is impossible to make a direct measurement and its value is obtained through relationships between other quantities. After the beginning of the pumping, it is occurred the lowering of water level inside the well. Then, the manometric height changes and as result the water flow also changes. The efficiency of a motor-pump set will also change along its useful life due to the equipment wearing, piping incrustations, leakages in the hydraulic system, obstructions of filters inside the well, closed or semi-closed valves, etc.

Therefore, converting all variables in (17) to meters, the most generic form of the *GEEI* is given by:

$$GEEI = \frac{P_{el}}{Q.H_T}$$

(18)

The *GEEI* defined in (18) can be used to analyze the well behavior along the time.

NEURAL APPROACH USED TO DETERMINE THE GLOBAL ENERGETIC EFFICIENCY INDICATOR

Among all necessary parameters to determine the proposed *GEEI*, the determination of the exploration flow is the most difficult to obtain in practice. The use of flow meters, as the electromagnetic ones, is very expensive. The use of rudimentary tests has provided imprecise results.

To overcome this practical problem, it is proposed here the use of artificial neural networks to determine the exploration flow from other parameters that have been measured before determining the *GEEI*.

Artificial Neural Networks (ANN) are dynamic systems that explore parallel and adaptive processing architectures. They consist of several simple processor elements with high degree of connectivity between them [4]. Each one of these elements is associated with a set of parameters, known as network weights, that allows the mapping of a set of known values (network inputs) to a set of associated values (network outputs).

The process of weight adjustment to suitable values (network training) is carried out through successive presentation of a set of training data. The objective of the training is the minimization between the output (response) generated by the network and the respective desired output. After training process, the network will be able to estimate values for the input set, which were not included in the training data.

In this work, an ANN will be used as a functional approximator, since the exploration flow of the well is a dependent variable of those ones that will be used as input variables. The functional approximation consists of mapping the relationship between the several variables that describe the behavior of a real system [5].

The ability of neural artificial networks to mapping complex nonlinear functions makes them an attractive tool to identify and to estimate models representing the dynamic behavior of engineering processes. This feature is particularly important when the relationship between several variables involved with the process is nonlinear and/or not very well defined, making its modeling difficult by conventional techniques.

A multilayer perceptron (MLP), as that shown in Figure 2, trained by the backpropagation algorithm, was used as a practical tool to determine the water flow from the measured parameters.

The input variables applied to the proposed neural network were the following:

- Level of water in meters (Ha) inside the well at the instant t.
- Manometric height in meters of water column (Hm) at the instant t.
- Electric power in Watts (Pel) absorbed from the electric system at the instant t.

The unique output variable was the exploration flow of the aquifer (Q), which is expressed in cubic meters per hour. It is important to observe that for each set of input values at a certain instant t, the neural network will return a result for the flow at that same instant t.

The determination of *GEEI* will be done by using in equation (18) the flow values obtained from the neural network and other parameters that come from experimental measurements.

To training of the neural network, all these variables (inputs and output) were measured and provided to the network. After training, the network was able to estimate the respective output variable. The values of the input variables and the respective output for a certain pumping period, which were used in the network training, are given by a set composed by 40 training patterns (or training vectors).

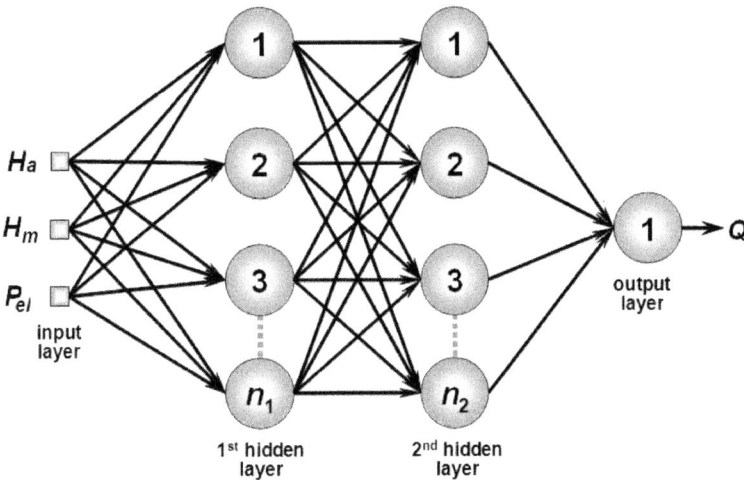

Figure 2. Multilayer perceptron used to determine the water flow.

These patterns were applied to a neural network of MLP type (Multilayer Perceptron) with two hidden layers, and its training was done using the *backpropagation* algorithm based on the Levenberg-Marquardt's method [6]. A description of the main steps of this algorithm is presented in the Appendix.

The network topology that was used is similar to that presented in Figure 2. The number of hidden layers and the number of neurons in each layer were determined from results obtained in [7,8]. The network is here composed by two hidden layers and the following parameters were used in the training process:

- Number of neurons of the 1st hidden layer: 15 neurons.
- Number of neurons of the 2nd hidden layer: 10 neurons.
- Training algorithm: Levenberg-Maquart.
- Number of training epochs: 5000 epochs.

At the end training process, the mean squared error obtained was 7.9×10^{-5}, which is a value considered acceptable for this application [7].

After training process, values of input variables were applied to the network and the respective values of flow were obtained in its output. These values were then compared with the measured ones in order to evaluate the obtained precision.

Table I shows some values of flow that were given by the artificial neural network (*QANN*) and those measured by experimental tests (*QET*).

Table 1. Comparison of results.

H_a (m)	H_m (m)	P_{el} (W)	Q_{ANN} (m³/h)	Q_{ET} (m³/h)
25.10	8.25	26,256	74.99	75.00
31.69	**40.50**	**26,155**	**53.00**	**62.00**
31.92	48.00	25,987	56.00	56.00
31.12	48.00	25,953	55.00	55.00
32.50	**48.00**	**25,970**	**54.08**	**54.00**
32.74	48.00	25,970	54.77	54.50
33.05	48.00	25,937	54.15	54.00
33.26	**48.00**	**25,954**	**58.54**	**54.00**
33.59	48.00	25,869	53.01	53.00
33.83	48.00	25,886	53.49	53.50
34.15	**48.00**	**25,887**	**53.50**	**53.00**
34.41	48.00	25,886	53.48	53.50
34.71	48.00	25,785	53.25	53.30
34.95	**48.00**	**25,870**	**53.14**	**53.00**
35.00	48.00	25,801	53.14	53.00

In this table, the values in bold were not presented to the neural network during the training.

When the patterns used in the training are presented again, it is noticed that the difference between the results is very small, reaching the maximum value of 0.35% of the measured value. When new patterns are used, the highest error reaches the value of 14.5%. It is also observed that the error value to new patterns decreases when they represent an operational stability situation of the motor-pump set, i.e., they are far away from the transitory period of pumping.

At this point, we should observe that it would be desirable a greater number of training patterns for the neural network, especially if it could be obtained from a wider variation of the range of values.

The proposed *GEEI* was determined by equation (18) and the measured values used were the electric power, the dynamic level, the geometric difference of level, the pressure of output in the well, and the water flow obtained from the neural network.

Figure 3 shows the behavior of *GEEI* during the analyzed pumping period.

The numeric values that have generated the graphic in Figure 3 are presented in Table 2.

Table 2. *GEEI* calculated using the artificial neural network.

Operation Time (min)	$GEEI_{(t)}$ $(Wh/m^3.m)$	Operation Time (min)	$GEEI_{(t)}$ $(Wh/m^3.m)$
0	7.420*	40	5.054
1	4.456*	45	5.139
2	5.738*	50	5.134
3	5.245*	55	5.115
4	4.896*	60	5.073
5	4.951*	75	5.066
6	4.689*	90	5.060
7	5.078*	105	5.042
8	4.840*	120	5.037
9	5.027*	135	5.042
10	5.090*	155	5.026
11	5.100*	185	5.032
12	5.092*	215	5.030
14	5.066*	245	5.040
16	5.044*	275	5.034
18	5.015*	305	5.027
20	5.006*	335	5.017
22	5.017	365	5.025
24	5.022	395	5.030
26	5.032	425	5.031
28	5.049	455	5.020
30	5.062	485	5.015
35	4.663		

* GEEI in transitory period (from 0 to 20 min of pumping).

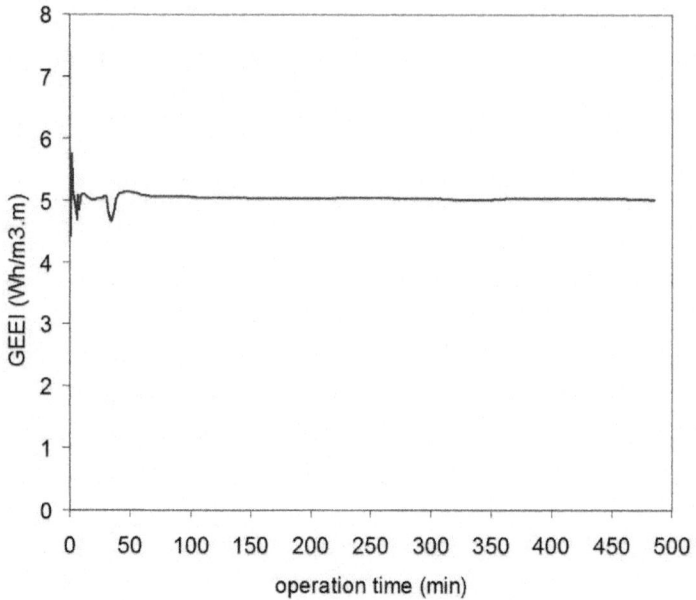

Figure 3. Behavior of the *GEEI* in relation to time.

ESTIMATION OF AQUIFER DYNAMIC BEHAVIOR US-ING NEURAL NETWORKS

In this section, artificial neural networks are now used to map the relationship between the variables associated with the identification process of aquifer dynamic behavior.

The general architecture of the neural system used in this application is shown in Figure 4, where two neural networks of type MLP, MLP-1 and MLP-2, constituted respectively by one and two hidden layers, compose this architecture.

The first network (ANN-1) has 10 neurons in the hidden layer and it is responsible by the computation of the aquifer operation level. The training data for ANN-1 were directly obtained from experimental measurements. It is important to note that this network has taken into account the present level and rest time of the aquifer.

The second network (ANN-2) is responsible by the computation of the aquifer dynamic level and it is composed by 2 hidden layers with both having 10 neurons. For this network, the training data were also obtained from experimental measurements. As observed in Figure 4, the ANN-1 output is provided as an input parameter to the ANN-2. Therefore, the computation of the aquifer dynamic level takes into account the aquifer operation level, the exploration flow and operation time.

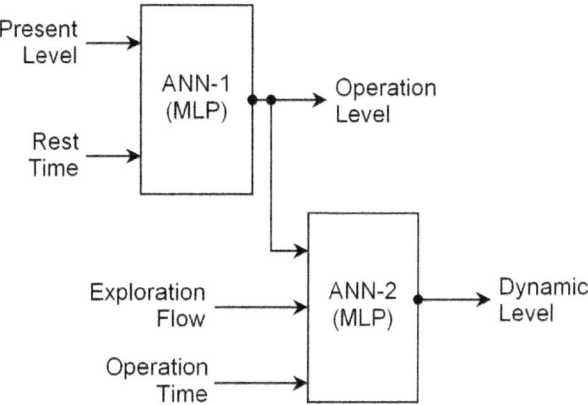

Figure 4. General architecture of the ANN used for estimation of aquifer dynamic behavior.

After training process of the neural networks, they were used for estimation of the aquifer dynamic level. The simulation results obtained by the networks are presented in Table 3 and Table 4.

Table 3. Simulation results (ANN-1).

Present Level (meters)	Rest Time (hours)	Operation Level (ANN-1)	Operation Level (Exact)	Relative Error (%)
115.55	4	103.59	104.03	0.43 %
125.86	9	104.08	104.03	0.05 %
141.26	9	105.69	104.03	1.58 %
137.41	8	102.95	104.03	1.05 %

Table 3 presents the simulation results obtained by the ANN-1 for a particular well. The operation levels computed by the network taking into account the present level and rest time of the aquifer were compared with those results obtained by measurements. In this table, the 'Relative Error' column provides the relative error between the values estimated by the network and those obtained by measurements.

Table 4. Simulation results (ANN-2).

Operation Flow (m³/h)	Operation Time (hours)	Dynamic Level (ANN-2)	Dynamic Level (Exact)	Relative Error (%)
145	14	115.50	115.55	0.04 %
160	2	116.10	116.14	0.03 %
170	6	118.20	117.59	0.52 %
220	21	141.30	141.26	0.03 %

The simulation results obtained by the ANN-2 are provided in Table 4. The dynamic level of the aquifer is estimated by the network in relation to operation level (computed by the ANN-1), exploration flow and operation time. These results are also compared with those obtained by measurements. In Table 4, the 'Relative Error' column gives the relative error between the values computed by the network and those from measurements.

These results show the efficiency of the neural approach used for estimation of aquifer dynamic behavior. The values estimated by the network are accurate to within 1.5% of the exact values for ANN-1 and 0.5 for ANN-2. From analysis of the results presented in Table 3 and 4, it is verified that the relative error between values provided by the network and those obtained by experimental measurements is very small. For ANN-1, the greatest relative error is 1.58 % (Table 3) and for ANN-2 is 0.52% (Table 4).

CONCLUSION

The management of systems that explore underground aquifers includes the analysis of two basic components: the water, which comes from the aquifer; and the electric energy, which is necessary to the transportation of the water to the consumption point or reservoir. Thus, the development of an efficiency indicator that shows the energetic behavior of a certain capitation system is of great importance to efficient management of the energy consumption, or still, to convert the obtained results in actions that become possible a reduction of energy consumption.

The obtained *GEEI* will indicate the global energetic behavior of the water capitation system from aquifers and will be an indicator of occurrences of abnormalities, such as tubing breaks or obstructions.

The application of the proposed methodology uses parameters that have easily been obtained in the water exploration system. The *GEEI* calculus can also be done by operators or to be implemented by means of computational system.

In addition, a novel methodology for estimation of aquifer dynamic behavior using artificial neural networks was also presented in this chapter. The estimation process is carried out by two feedforward neural networks. Simulation results confirm that proposed approach can be efficiently used in these types of problem. From results, it is possible to simulate several situations in order to define appropriate management plans and policies to the aquifer.

The main advantages in using this neural network approach are the following: i) velocity: the estimation of dynamic levels are instantly computed and it is appropriated for application in real time, ii) economy and simplicity: reduction of operational costs and measurement devices, and iii) precision: the values estimated by the proposed approach are as good as those obtained by physical measurements.

APPENDIX

The mathematic model that describes the behavior of the artificial neuron is expressed by the following equation:

$$u = \sum_{i=1}^{n} w_i \cdot x_i + b$$

(19)

$$y = g(u)$$

(20)

where n is the number of inputs of the neuron; xi is the i-th input of the neuron; wi is the weight associated with the i-th input; b is the threshold associated with the neuron; u is the activation potential; $g(\ )$ is the activation function of the neuron; y is the output of the neuron.

Basically, an artificial neuron works as follows:

- Signals are presented to the inputs.
- Each signal is multiplied by a weight that represents its influence in that unit.
- A weighted sum of the signals is made, resulting in a level of activity.
- If this level of activity exceeds a certain threshold, the unit produces an output.

To approximate any continuous nonlinear function a neural network with only a hidden layer can be used. However, to approximate non-continuous functions in its domain it is necessary to increase the amount of hidden layers. Therefore, the networks are of great importance in mapping nonlinear processes and in identifying the relationship between the variables of these systems, which are generally difficult to obtain by conventional techniques.

The network weights (wj) associated with the j-th output neuron are adjusted by computing the error signal linked to the k-th iteration or k-th input vector (training example). This error signal is provided by:

$$e_j(k) = d_j(k) - y_j(k)$$

(21)

where $dj(k)$ is the desired response to the j-th output neuron.

Adding all squared errors produced by the output neurons of the network with respect to k-th iteration, we have:

$$E(k) = \frac{1}{2} \sum_{j=1}^{p} e_j^2(k)$$

(22)

where p is the number of output neurons.

For an optimum weight configuration, $E(k)$ is minimized with respect to the synaptic weight wji. The weights associated with the output layer of the network are therefore updated using the following relationship:

$$w_{ji}(k + 1) \leftarrow w_{ji}(k) - \eta \frac{\partial E(k)}{\partial w_{ji}(k)}$$

(23)

where wji is the weight connecting the j-th neuron of the output layer to the i-th neuron of the previous layer, and η is a constant that determines the learning rate of the backpropagation algorithm.

The adjustment of weights belonging to the hidden layers of the network is carried out in an analogous way. The necessary basic steps for adjusting the weights associated with the hidden neurons can be found in [4].

Since the backpropagation learning algorithm was first popularized, there has been considerable research into methods to accelerate the convergence of the algorithm.

While backpropagation is a steepest descent algorithm, the Marquardt-Levenberg algorithm is similar to the quasi-Newton method, which was designed to approach second-order training speed without having to compute the Hessian matrix.

When the performance function has the form of a sum of squared errors like that presented in (22), then the Hessian matrix can be approximated as

$$H = J^T \cdot J$$

(24)

and the gradient can be computed as

$$g = J^T \cdot e \tag{25}$$

where e is a vector of network errors, and J is the Jacobean matrix that contains first derivatives of the network errors with respect to the weights and biases.

The Levenberg-Marquardt algorithm uses this approximation to the Hessian matrix in the following Newton-like update:

$$w(k+1) \leftarrow w(k) - (J^T \cdot J + \mu \cdot I)^{-1} \cdot J^T \cdot e \tag{26}$$

When the scalar μ is zero, this is Newton›s method, using the approximate Hessian matrix. When μ is large, this produces a gradient descent with a small step size. Newton's method is faster and more accurate near to an error minimum, so the aim is to shift toward Newton's method as quickly as possible.

Thus, μ is decreased after each successful step (reduction in performance function) and is increased only when a tentative step would increase the performance function. In this way, the performance function is always reduced at each iteration of the algorithm [6].

This algorithm appears to be the fastest method for training moderate-sized feedforward neural networks (up to several hundred weights).

REFERENCES

1. P. A. Domenico, 2011 Concepts and Models in Groundwater Hydrology. New York: McGraw-Hill.

2. P. A. Domenico, F. W. Schwartz, 1990 Physical and Chemical Hydrogeology. New York: John Wiley and Sons.

3. N. J. Saggioro, 2001 Development of Methodology for Determination of Global Energy Efficiency Indicator to Deep Wells. Master's degree dissertation (in Portuguese). São Paulo State University.

4. S. Haykin, 2008 Neural Networks and Learning Machines. New York: Prentice-Hall, 3rd edition.

5. M. Anthony, P. L. Barlett, 2009 Neural Network Learning: Theoretical Foundations. Cambridge: Cambridge University Press.

6. M. T. Hagan, M. B. Menhaj, 1994 Training Feedforward Networks with the Marquardt Algorithm IEEE Transactions on Neural Networks 5 6 989 993

7. I. N. Silva, N. J. Saggioro, J. A. Cagnon, 2000 Using neural networks for estimation of aquifer dynamical behavior In: proceedings of the International Joint conference on Neural Networks, IJCNN2000 24-27

July 2000, Como, Italy

8.   J. A. Cagnon, N. J. Saggioro, I. N. Silva, 2000 Application of neural networks for analysis of the groundwater aquifer behavior In: Proceedings of the IEEE Industry Applications Conference, INDUSCON2000 06-09 November, Porto Alegre, Brazil

9.   F. G. Driscoll, 1986 Groundwater and Wells. Minneapolis: Johnson Division.

10. D. M. Allen, N. Schuurman, Q. Zhang, Using Fuzzy Logic for Modeling Aquifer Architecture Journal of Geographical Systems 2007 9 289 310

11. J. P. Delhomme, 1989 Spatial Variability and Uncertainty in Groundwater Flow Parameters: A Geostatistical Approach Water Resources Research 15 2 269 280

12. K. Koike, H. Sakamoto, M. Ohmi, Detection and Hydrologic Modeling of Aquifers in Unconsolidated Alluvial Plains though Combination of Borehole Data Sets: A Case Study of the Arao Area, Southwest Japan Engineering Geology 2001 62 4 301 317

13. J. Scibek, D. M. Allen, 2006 Modeled Impacts of Predicted Climate Change on Recharge and Groundwater Levels Water Resources Research 42 18 doi: 10.1029/ 2005WR004742

14. S. Fu, Y. Xue, 2011 Identifying aquifer parameters based on the algorithm of simple pure shape In: Proceedings of the International Symposium on Water Resource and Environmental Protection, ISWREP2011 20-22 May, Xi'an, China

15. G. Jinyan, L. Yudong, M. Yuan, H. Mingchao, L. Yan, L. Hongjuan, 2011 A mathematic time dependent boundary model for flow to a well in an unconfined aquifer In: Proceedings of the International Symposium on Water Resource and Environmental Protection, ISWREP2011 20-22 May 2011, Xi'an, China

16. Z. Hongfei, G. Jianqing, 2010 A mathematic time dependent boundary model for flow to a well in an unconfined aquifer In: Proceedings of the 5th International Conference on Computer Sciences and Convergence Information Technology, ICCIT2010 30 November to 02 December 2010, Seoul, Korea

17. E. Cameron, G. F. Peloso, An Application of Fuzzy Logic to the Assessment of Aquifers' Pollution Potential Environmental Geology 2001 40 11-12 1305 1315

18. A. Gemitzi, C. Petalas, V. A. Tsihrintzis, V. Pisinaras, Assessment of Groundwater Vulnerability to Pollution: A Combination of GIS, Fuzzy

Logic and Decision Making Techniques Environmental Geology 2006 49 5 653 673

19. Y. S. Hong, M. R. Rosen, R. R. Reeves, 2002 Dynamic Fuzzy Modeling of Storm Water Infiltration in Urban Fractured Aquifers Journal of Hydrologic Engineering 7 5 380 391

20. X. He, J. J. Liu, 2009 Aquifer parameter identification with ant colony optimization algorithm In: Proceedings of the International Workshop on Intelligent Systems and Applications, ISA2009 23-24 May, Wuhan, China

Chapter 7

MODELING URBAN HYDROLOGY: A COMPARISON OF NEW URBANIST AND TRADITIONAL NEIGHBORHOOD DESIGN SURFACE RUNOFF

Christopher Andrew Day[1], and Keith Allen Bremer[2]

[1]Department of Geography and Geosciences, University of Louisville, Louisville, USA

[2]Department of Geography, Texas State University-San Marcos, San Marcos, USA.

ABSTRACT

Urban development generally leads to an increase in impervious cover resulting in a greater volume of surface runoff following storm activity. However, the type of urban development in place strongly controls the degree of impervious cover generated. Traditional neighborhood designs focus on a medium-to-low urban density spread over larger areas, while new urbanist neighborhood designs incorporate more diversity by increasing urban density across smaller areas. The purpose of this study is to model and compare the potential surface runoff for two urban neighborhoods in Austin, Texas-Circle C Ranch, a traditional neighborhood design, and Mueller, a new urbanist development for a 10-year 24-hour storm scenario. Potential surface runoff was calculated by layering various geospatial datasets representing the physical characteristics of both study sites within the Watershed Modeling System (WMS) to configure the HEC-HMS runoff model. Results initially imply that the higher density new urbanist neighborhood significantly increases total and peak storm runoff compared to the traditional neighborhood. However, a greater number of residential units are available at Mueller over the same area as Circle C Ranch. When taking this into account the increased potential surface runoff is negated at the new urbanist site. Although new urbanist neighborhoods will usually contain more residential units than traditional developments when compared at the same scale, the higher urban density associated with these neighborhoods demand the development of more effective stormwater retention systems to cope with a potential increase in surface runoff.

INTRODUCTION

Urban development affects the amount of potential surface runoff generated during storms by changing the amount of impervious cover across the landscape [1-3]. In addition to increasing surface runoff, urban development also modifies the volume of groundwater recharge, lowers water tables, increases peak discharge, and decreases base flow during drier periods [4,5]. These modifications depend on the type of urban development in place.

New urbanism is a type of sustainable development that is designed to reduce automobile use, increase walking and cycling, and increase the diversity of land uses while incorporating traditional and new practices of planning at all scales [6]. Moreover, new urbanism is a type of low impact development (LID) that contains elements such as cluster development and bio-retention. LID can mitigate problems associated with storm water runoff by increasing resilience and utilizing best management practices [7,8]. Traditional neighborhood development (TND) is limited to the neighborhood scale and incorporates traditional planning practices such as large lot and single family zoning [9]. TND are not considered as LID unless further steps have been taken to implement specific LID features. New urbanism is touted as a more environmentally sustainable development than TND, which will typically contain greater amounts of impervious cover [10].

While research implies that LID does often reduce total stormwater runoff and increase the runoff lag time when compared to TND [11-13], more research needs to be carried out which compare neighborhoods of similar size and scale in order to make further accurate assessments of LID and their impact on stormwater runoff. Several obstacles pertinent to stormwater runoff have been noted concerning LID planning. Many current zoning and regulatory statutes can hinder the implementation of LID concepts and philosophies [14]. These features include minimum street width for public services, concrete curbs and gutters, the absence of runoff collection ponds due to public health concerns, and other elements that may not fit into the visually pleasing aesthetic design [14]. As a result, a comparison of three urban neighborhoods ranging from high-to-low density actually found that the medium density neighborhood displayed the longest peak run off lag times due to more effective us age of stormwater retention systems [15].

An increase in geospatial and modeling capability has increased the opportunity of analyzing urban development impacts on stormwater runoff in recent years. Remote sensing data coupled with geographic information science (GIS) systems and runoff modeling software have been used more frequently to study the interaction between rainfall events and urban surfaces leading to runoff [16-18]. The purpose of this research is to utilize these techniques to model and compare the potential surface runoff for two similar-sized new urbanist and traditional neighborhoods in Austin, Texas.

STUDY AREA

The study area includes two neighborhoods, one new urbanist, and one traditionally developed neighborhood in Austin, Texas **(Figure 1)**. Austin-Mueller (Mueller) is a new urbanist neighborhood located in north-central Austin approximately three miles from downtown Austin on the site of the city's old Robert Mueller airport. Mueller is the most recent master-planned community in Austin that focuses on new urbanism as a vehicle for sustainability including a mixture of home types, sizes, and price ranges. Circle C Ranch is a traditional neighborhood development that originated in the late 1980s. The neighborhood contains mostly single-family homes that are situated on medium to large lots with traditional planning practices in place [19].

Regarding physical characteristics that may impact stormwater runoff, Austin receives, on average, 870 mm precipitation annually [20]. The majority of this total occurs in the months of April and May when violent storms develop from Pacific cold fronts moving rapidly across the south-central Texas region, resulting in severe flooding [21]. Another important factor concerning runoff is the soil which heavily controls the amount of infiltration-to-surface-runoff ratio during storm events. Soils may be classified into one of four hydrologic groups (A, B, C, D) that reflect their drainage capability. Group A soils are characterized by high infiltration rates to give low runoff potential following precipitation, while group D soils have low infiltration rates to increase runoff potential [22]. Soil coverage across both sites is typical of the south-central Texas region.

Mueller is dominated by the Lewisville and Altoga series soils which range from well-to-moderately drained silty-clay soils underlain by fractured chalk or limestone, classified in the B-C soil hydrologic groups. Smaller instances of the Houston Black and Patrick soil series are also present at these sites classified into the moderately-to-poorly drained B-D soil hydrologic grouping. At Circle C Ranch, the Tarrant soil series dominates as a stony-clay type soil (hydrologic group C) with the moderately well-drained (C group) Speck series present to the south and west of the site [22].

METHODS

The methodology workflow incorporates a series of geospatial data sources and techniques in order to calculate potential surface runoff at both study areas **(Figure 2)**. Land use/cover data were obtained for both sites from 1m resolution Digital Orthophoto Quarter Quad (DOQQ) images from 2010. In order to directly compare the runoff generated between the two sites, the larger Circle C Ranch site was trimmed down to match the area of Mueller, using road boundaries within the sub -division as the new boundaries for Circle C Ranch. This gave two images covering an equal area of 0.7 km2 with the Mueller site containing 751 residential units and Circle C Ranch 511.

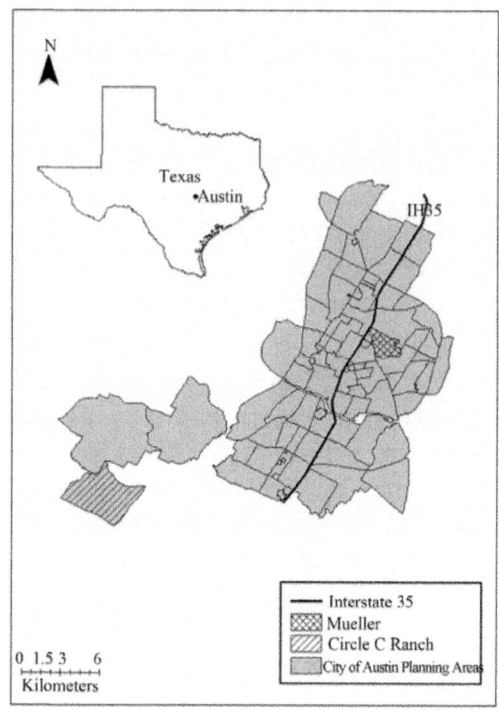

Figure 1. Study are as within Austin, Texas

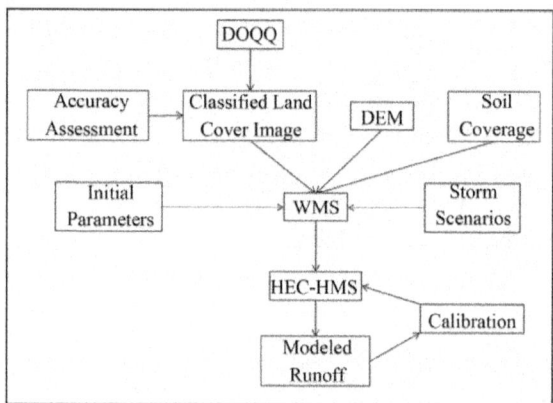

Figure 2. Methodology workflow.

The imagery was initially loaded in ArcMap before performing a supervised classification technique using the maximum likelihood algorithm. Following a visual inspection of the images, four land cover classes were identified as urban/impervious, forest, grass, and surface water (**Figure 3**). The classification accuracy was verified by rechecking the classified images with the original imagery. The classified images were then loaded into the Watershed Modeling System (WMS) software and combined with a digital elevation model (DEM) to calculate slope and hydraulic length (the longest flow path across each site, L) for both sites. Finally, soil coverage's, containing the soil hydrologic groups for the soils at both sites, from the State Soil Geographic Database (STATSGO) were loaded into the model in order to calculate infiltration losses during storm activity, similar to previous research techniques [23] (**Figure 4**).

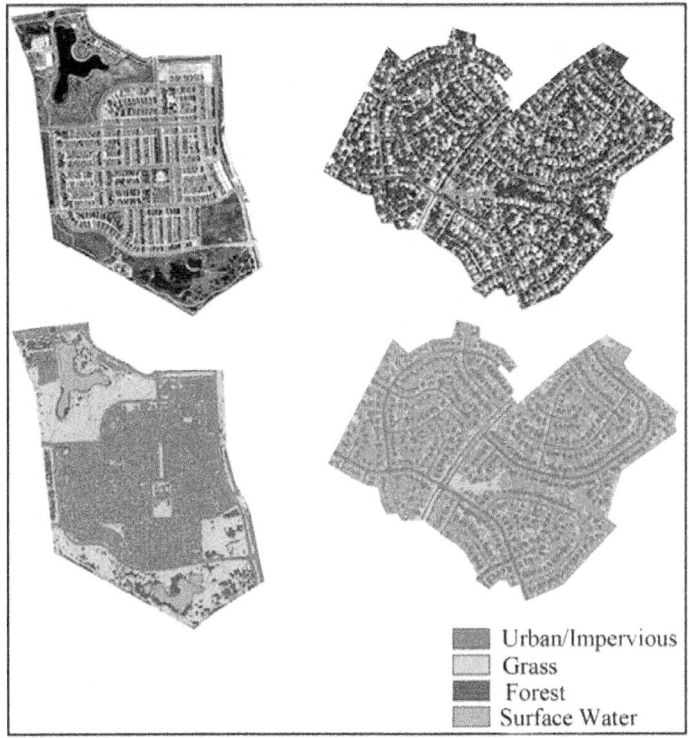

Figure 3. Land cover classification from DOQQ imagery for Mueller (left) and Circle C Ranch (right).

Surface runoff was calculated using the HEC-HMS model for a 10-year 24 hour storm scenario based on the surface and soil hydrologic group cover for each site. The HEC-HMS model was originally developed by the US Army Corps of Engineers (US ACE) as a lumped parameter model, capable of routing surface flow into a series of drainage basins towards an outlet [23,24]. Various methods are available within HEC-HMS to determine runoff versus infiltration. The Soil Conservation Service (SCS) method was chosen for this study based on its success at modeling surface runoff in other urban runoff studies [18,25,26], and the availability of the necessary physical data at both study sites in Austin. It is also ideally suited for modeling drainage areas of less than 2000 acres (~8 km2) [27]

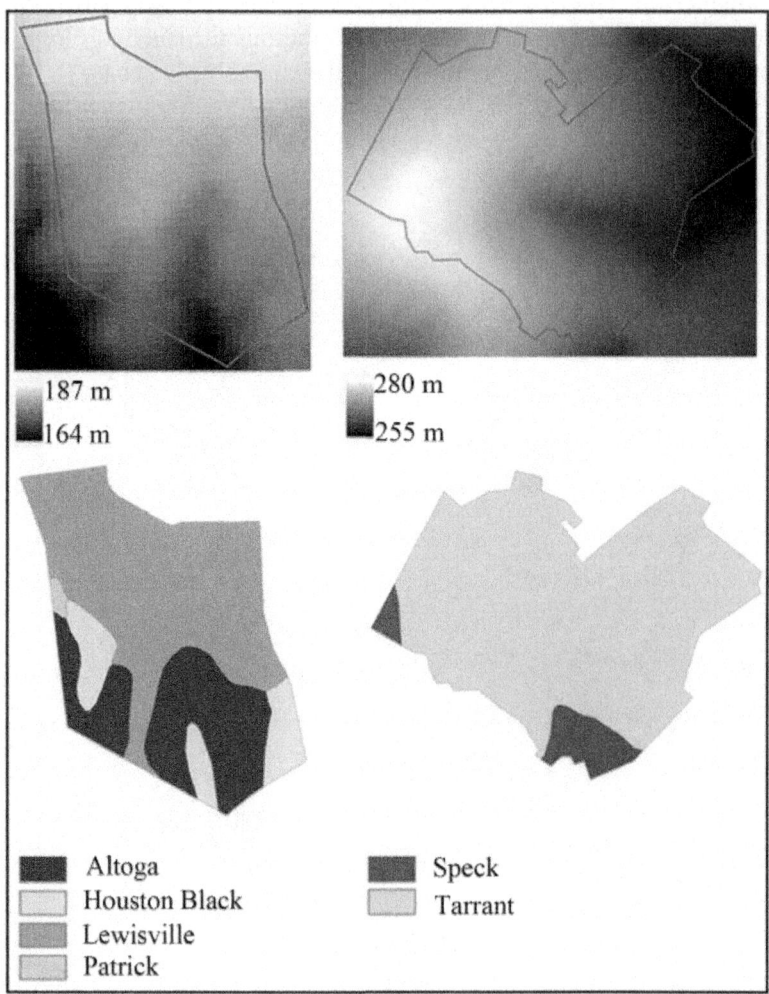

Figure 4. DEM and soil coverage for Mueller (left) and Circle C Ranch (right).

The SCS method calculates initial precipitation losses (the initial abstraction) and ultimately the volume of water available for surface runoff based on soil permeability and land cover by prescribing a predetermined "curve number" to each surface and soil hydrologic group cover (Equations (1) and (2)).

$$Q = \frac{(P - Ia)^2}{P - Ia + S} \tag{1}$$

Q = runoff depth;

P = 24-hour storm precipitation depth;

Ia = initial abstraction (0.2S);

S = infiltration/retention losses (Equation (2))

$$S = \left(\frac{1000}{CN}\right) - 10 \tag{2}$$

CN = curve number for areal soil and land cover.

Higher curve numbers result from land cover and soil hydrologic groups that allow decreased infiltration, resulting in a greater volume of water made available for surface runoff. By overlaying the classified land cover data with the soil hydrologic group coverage data, a composite curve number could be generated for each site (Equation (3)) [24] .

$$CN_{comp} = \frac{\sum A_i CN_i}{\sum A_i} \tag{3}$$

CN_{comp} = composite curve number;

A_i = drainage area of each area with uniform land and

soil coverage;

CN_i = curve number of each A_i.

The curve numbers used in Equations (2) and (3) for soil hydrologic groups and various land cover surfaces are given in **Table 1**.

Runoff volumes were then generated to produce hydrographs which determined the peak runoff in cubic meters per second (cms) and lag time between peak precipitation and runoff at each site. The 24-hour storm precipitation depth in equation 1 was taken from a 10-year 24 hour storm scenario for the Austin area based on the availability of local historical hydrological data for model calibration later (**Table 2**). Within the WMS modeling software the SCS method initially estimates basin lag time using the physical basin parameters in Equation (4), (**Table 3**):

$$T_{lag} = L^{0.8} \frac{(S+1)^{0.7}}{1900\sqrt{Y}}$$

(4)

T_{lag} = basin lag time;

L = hydraulic length;

S = infiltration/retention losses (Equation (2));

Y = mean slope.

Calibration of the HEC-HMS model is normally achieved by comparing the modeled runoff with observed runoff obtained from a US Geological Survey stream gauge at the outlet of the modeled catchment site [23,29].

Table 1. Example runoff curve numbers for various land covers by soil hydrologic group [27].

Land Cover	Soil Hydrologic Group			
	A	B	C	D
Impervious Surfaces	98	98	98	98
Woods/Forest	30	55	70	77
Grass	39	61	74	80
Surface Water	0	0	0	0

Table 2. Approximate precipitation depths for a 10-year 24-hour storm in the Austin area [28].

Time period	Precipitation depth (mm)
15 min	35.6
1 hour	68.6
2 hours	86.4
3 hours	94.0
6 hours	109.2
12 hours	121.9
24 hours	152.4

Table 3. SCS model parameters gene rated by WM

Site	Hydraulic length, L (m)[a]	Infiltration losses, S	Slope, Y (%)	Basin lag time, T_{lag} (hr)
Mueller	994	1.6	1.8	0.5
Circle C Ranch	1020	3.2	2.3	0.62

[a]Although meters are given, the equation requires L input in feet.

This was not directly possible as neither sit contained an active stream gauge for model comparison located at the site outlets. To account for this, calibration of the runoff model took place by comparing the peak flow generated from a 10-year 24-hour storm with the observed peak flow from the nearest active stream gauge located approximately 2.4 km downstream from the Mueller site (Boggy Creek USGS# 08 158035). In this case the model ran using the initial conditions calculated by HECHMS from the physical site data, before adjusting the key parameter, initial abstraction, to match the proportional observed peak runoff generated at Boggy Creek. This took into account the larger catchment area of the Boggy Creek gauge location. Initial peak runoff was overestimated, and subsequent lag times underestimated, as a result of low initial abstraction parameter values generated by the model. This was corrected by increasing the initial abstraction value until the peak runoff value at Mueller proportionally matched the value at the Boggy Creek site, similar to the approach adopted by previous urban runoff modeling research [23,29]. Adjustment of the initial abstraction value for Circle C Ranch followed based on the lower CN value for that site (**Table 4**).

RESULTS

The Mueller site contained a much greater proportion of urban/impervious cover, totaling 50% compared to the Circle C Ranch coverage of 36% (**Figure 3, Table 5**). The impervious area of the Mueller neighborhood is also clustered around a central area, surrounded by non-impervious surfaces, which typifies new urbanist developments.

The Circle C Ranch site displays a more uniform spread of all surfaces, with impervious surfaces distributed across the entire site. While Mueller does display 17% more grass coverage, the majority of the Circle C Ranch site is covered in forest, totaling 51% compared to Mueller's 16%. Mueller also includes 4% surface water coverage in the form of two ponds located to the south and northwest of the site.

Regarding runoff, initially the two hydrographs produced by the model appear similar, but closer inspection reveals three key differences between Mueller and Circle C Ranch in response to the 10-year storm scenario (**Figure 5**). Firstly, the peak runoff increased by 64% from 0.99cms at Circle C Ranch to 1.55 cm at the Mueller site. Secondly, the storm lag time displayed a lower value by 31 minutes at Mueller, which equated to a 59% decrease in time from Circle C Ranch storm response. Lastly, the runoff coefficient (proportion of rainfall to runoff), increased by 5.9% at Mueller, again highlighting that a greater proportion of rainfall during the storm becomes surface runoff at this location. The results suggest that the new urbanist site at Mueller actually generates the greater volume of stormwater runoff (42,000 m³ vs. 35,700 m³ at Circle C Ranch).

Table 4. Curve numbers and initial abstraction values used in model.

Site	Default Initial Abstraction (Ia)	Calibrated Initial Abstraction (Ia)	Curve Number
Mueller	0.2	0.26	86
Circle C Ranch	0.2	0.32	78

Table 5. Proportion of surface cover at Circle C Ranch and Mueller sites

Surface cover	Circle C Ranch	Mueller
Impervious	36%	50%
Forest/Woods	51%	16%
Grass	13%	30%
Water	0%	4%

Furthermore, with both sites displaying similar physical properties in terms of area, relief, hydraulic length and soil hydrologic group characteristics, the greater extent of impervious surface coverage compared to the traditional site at Circle C Ranch is chiefly responsible for this.

However, it must be addressed that new urbanist developments focus on clustered development practices that have a higher density of residential development than a traditional urban development practice over a similar area. In this case Mueller contains 751 residential units compared to Circle C Ranch's 511, a total difference of 240 units over the 0.7 km2 area. Taking this into account Circle C Ranch would theoretically generate a greater volume of runoff at 69,863 m3 per 1000 units vs. 55,925 m3 per 1000 units at Mueller, a difference of just under 14,000 m3. As a result Circle C Ranch and other similar traditional urban developments, taken as a whole, will likely generate a greater volume of surface runoff than their new urbanist counterparts in terms of their total footprint on the landscape.

Of further note are the landscaped retention systems in place at the Mueller site which are designed to limit the effects of stormwater runoff, practices that are often not included across traditional developments. Bio-retention ponds are key features of new urbanist developments which aim to capture and store excess runoff following storm events. Mueller has two such ponds in place, to the north and south which have been aesthetically landscaped into the development blueprint. The DEM datasets used in this study do not capture any of these largescale landscaping changes implemented at the Mueller site, assuming that the majority of stormwater runoff will follow the original topography and drainage patterns. However the purpose of this paper was to investigate the potential surface runoff generated from this kind of development in comparison to a traditional neighborhood. The fact that new urbanist sites will often cluster their development in a bid to reduce the overall footprint of the site means that without these kinds of retention systems in place a greater volume of runoff could potentially be generated and lag times reduced following storm events as seen in this study.

CONCLUSIONS

A modeling framework has been developed to analyze the impacts of urban neighborhood design on storm runoff for the city of Austin, Texas. By layering a series of datasets that represent the physical landscape (land cover, soil, and relief) within the Watershed Modeling System (WMS) the HEC-HMS runoff model has generated peak runoff and storm lag times for a new urbanist and traditional neighborhood. The results imply that when directly comparing these types of urban design on a similar scale, the new urbanist neighborhood has the propensity to generate larger peak flows and shorter lag times as a function of the high density urban footprint associated with this type of neighborhood.

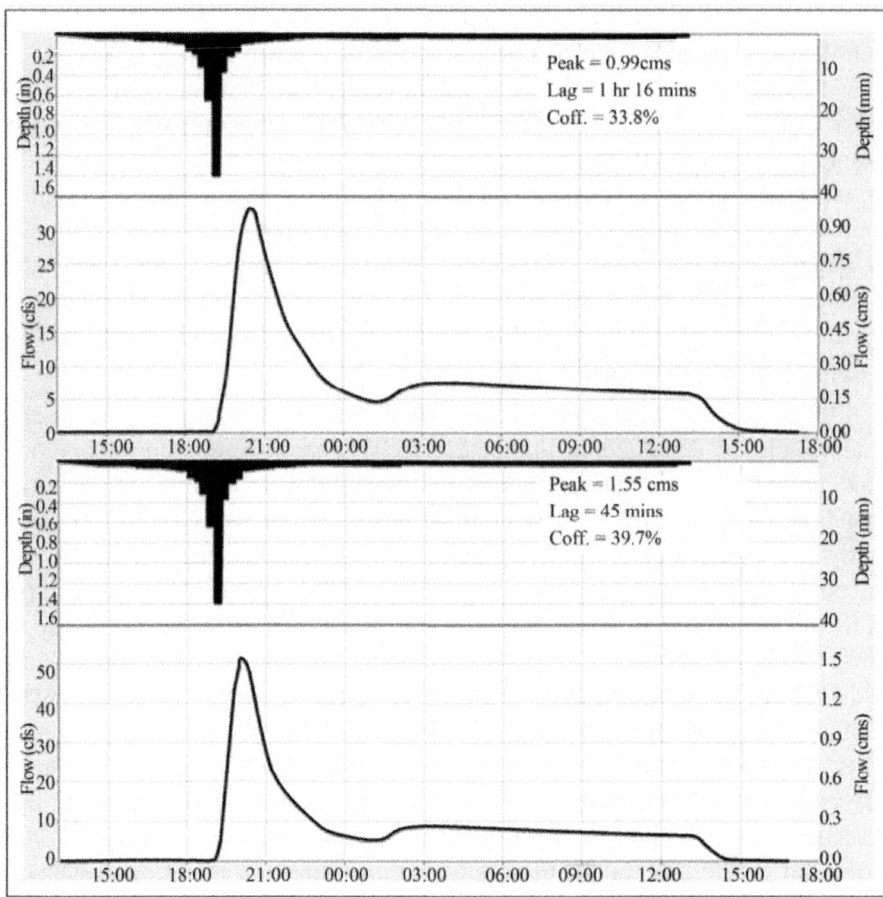

Figure 5. Modeled runoff hydrographs for Circle C Ranch (top) and Mueller (bottom).

Consequently it is imperative that flood retention or reduction measures are included in these neighborhood designs in order to mitigate the impacts of potential flooding both within and surrounding these new urbanist neighborhoods. Furthermore, while new urbanist neighborhoods have LID elements designed within them to reduce runoff and pollutants at a larger scale these results suggest more research is needed to determine how well, at the smaller scale, these elements work with other neighborhood designs and to what level they reduce or increase pollutant runoff.

The methodology employed in this research demonstrates the potential of combining and manipulating a series of datasets within GIS and modeling software to ascertain the potential surface runoff generated within urban areas at the sub-drainage basin scale. However further research should also be conducted that compares potential runoff output from infiltration abstraction methods other than the SCS method as employed in this research. Also with the increase in development of new urbanist neighborhood s within US cities, similar research may be conducted that compares the potential runoff between these neighborhoods. Their non-traditional development and design often makes them unique from one another and thus could generate significantly different runoff outputs from similar storm scenarios

REFERENCES

1. D. B. Booth, D. Hartley and R. Jackson, "Forest Cover, Impervious-Surface Area, and the Mitigation of Stormwater Impacts," Journal of the American Water Resources Association, Vol. 38, No. 2, 2002, pp. 835-845. doi:10.1111/j.1752-1688.2002.tb01000.x

2. L. B. Leopold, "Hydrology for Urban Planning—A Guidebook on the Hydrological Effects of Urban Land Use," US Geological Survey, Washington DC, 1968.

3. C. J. Walsh, A. H. Roy, J. W. Feminella, P. D. Cottinghan, P. M. Groffman and R. P. Morgan, "The Urban Stream Syndrome: Current Knowledge and the Search for a Cure," Journal of the North American Benthological Society, Vol. 24, No. 3, 2005, pp. 706-723. doi:10.1899/04-020.1

4. K. Gilroy and R. McCuen "Spatio-Temporal Effects of Low Impact Development Practices," Journal of Hydrology, Vol. 367, No. 3-4, 2009, pp. 228-236. doi:10.1016/j.jhydrol.2009.01.008

5. P. M. Groffman, N. J. Boulware, W. C. Zipperer, R. V. Pouyat, L. E. Band and M. F. Colosimo, "Soil Nitrogen Cycle Processes in Urban Riparian Zones," Environmental Science & Technology, Vol. 36, No. 21, 2002, pp. 4547-4552. doi:10.1021/es020649z

6. J. Dill, "Evaluating a New Urbanist Neighborhood," Berkeley Planning Journal, Vol. 19, No. 1, 2006, pp. 59-78.

7. E. Bedan and J. Clausen, "Stormwater Runoff Quality and Quantity from Traditional and Low Impact Development watersheds," Journal of the American Water Resources Association, Vol. 45, No. 4, 2009, pp. 998-1008. doi:10.1111/j.1752-1688.2009.00342.x

8. C. Pyke, M. Warren, T. Johnson, J. LaGro, J. Scharfenderg, P. Groth, R. Freed, W. Schroeer and E. Main, "Assessment of Low Impact Development for Managing Stormwater with Changing Precipitation Due to Climate Change," Landscape and Urban Planning, Vol. 103, No. 2, 2011, pp. 166-173. doi:10.1016/j.landurbplan.2011.07.006

9.   O. J. Furuseth, "Neotraditional Planning: A New Strategy for Building Neighborhoods?" Land Use Policy, Vol. 14, No. 3, 1997, pp. 201-213. doi:10.1016/S0264-8377(97)00002-1

10. P. R. Berke, J. MacDonald, N. White, M. Holmes, D. Line, K. Oury and R. Ryznar, "Greening Development to Protect Watersheds: Does New Urbanism make a Difference?" Journal of the American Planning Association, Vol. 69, No. 4, 2003, pp. 397-413. doi:10.1080/01944360308976327

11. M. E. Dietz, "Low Impact Development Practices: A Review of Current Research and Recommendations for Future Directions," Water, Air, and Soil Pollution, Vol. 186, No. 1-4, 2007, pp. 351-363. doi:10.1007/s11270-007-9484-z

12. J. K. Holman-Dodds, A. A. Bradley and K. W. Potter, "Evaluation of Hydrologic Benefits of Infiltration Based Urban Stormwater Management," Journal of the American Water Resources Association, Vol. 39, No. 1, 2003, pp. 205-215. doi:10.1111/j.1752-1688.2003.tb01572.x

13. M. J. Hood, J. C. Clausen and G. S. Warner, "Comparison of Stormwater Lag Times for Low Impact and Traditional Residential Development," Journal of the American Water Resources Association, Vol. 43, No. 4, 2007, pp. 1036-1046. doi:10.1111/j.1752-1688.2007.00085.x

14. P. A. Davis, "Green Engineering Principals Promote Low-Impact Development," Environmental and Science Technology, Vol. 39, No. 16, 2005, pp. 338A-344A. doi:10.1021/es053327e

15. D. Burns, T. Vitvar, J. McDonnell, J. Hassett, J. Duncan and C. Kendall, "Effects of Suburban Development on Runoff Generation in the Croton River Basin, New York, USA," Journal of Hydrology, Vol. 311, No. 1-4, 2005, pp. 266-281. doi:10.1016/j.jhydrol.2005.01.022

16. S. D. Khan, "Urban Development and Flooding in Houston Texas, Inferences from Remote Sensing Data using Neural Network Technique," Environmental Geology, Vol. 47, No. 8, 2005, pp. 1120-1127. doi:10.1007/s00254-005-1246-x

17. Y. P. Lin, Y. B. Lin, Y. T. Wang and N. M. Hong, "Monitoring and Predicting Land-Use Changes and the Hydrology of the Urbanized Paochiao Watershed in Taiwan Using Remote Sensing Data, Urban Growth Models and a Hydrological Model," Sensors, Vol. 8, No. 2, 2008, pp. 658-680. doi:10.3390/s8020658

18. S. Suriya and B. V. Mudgal, "Impact of Urbanization on Flooding: The Thirusoolam Sub Watershed—A Case Study," Journal of Hydrology, Vol.

412-413, 2012, pp. 210-219. doi:10.1016/j.jhydrol.2011.05.008

19. Circle C Ranch, "Circle C Ranch Homeowners Association," 2012. http://www.circlecranch.info

20. National Oceanographic and Atmospheric Administration (NOAA), "National Weather Service Forecast Office, Austin/San Antonio, TX," 2012. http://www.nws.noaa.gov/climate/xmacis.php?wfo=ewx

21. R. A. Earl and T. Kimmel, "Means and Extremes: The Weather and Climate of South-central Texas," In: J. F. Petersen and J. A. Tuason, Eds., A Geographic Glimpse of Central Texas and the Borderlands, National Council for Geographic Education, Pennsylvania, 1995, pp. 31-40.

22. US Soil Conservation Service (US SCS), "Soil Survey of Travis County, Texas," US SCS, Austin, 1974. doi:10.1016/j.jenvman.2006.06.023

23. C. McColl and G. Aggett, "Land Use Forecasting and Hydrologic Model Integration for Improved Land Use Decision Support," Journal of Environmental Management, Vol. 84, No. 4, 2007, pp. 494-512.

24. US Army Corps of Engineers (US ACE), "Hydrologic Modeling System HEC-HMS: Technical Reference Manual," US ACE, Washington DC, 2000.

25. Y. Guo and M. Markus, "Analytical Probabilistic Approach for Estimating Design Flood Peaks of Small Watersheds," Journal of Hydrologic Engineering, Vol. 16, No. 11, 2011, pp. 847-857. doi:10.1061/(ASCE) HE.1943-5584.0000380

26. C. J. Woltemade, "Impact of Residential Soil Disturbance on Infiltration Rate and Stormwater Runoff," Journal of the American Water Resources Association, Vol. 46, No. 4, 2010, pp. 700-711. doi:10.1111/j.1752-1688.2010.00442.x

27. US Soil Conservation Service (US SCS), "National Engineering Handbook, Section 4, Hydrology," US SCS, Washington DC, 1972.

28. W. H. Asquith and M. C. Roussel, "Atlas of Depth-Duration Frequency of Precipitation and Annual Maxima for Texas," Scientific Investigations Report 2004-5041, US Geological Survey, Austin, 2004.

29. M. R. Knebl, Z. L. Yang, K. Hutchinson and D. R. Maidment, "Regional Scale Flood Modeling Using NEXRAD Rainfall, GIS, and HEC-HMS/ RAS: A Case Study for the San Antonio River Basin Summer 2002 Storm Event," Journal of Environmental Management, Vol. 75, No. 4, 2005, pp. 325-336. doi:10.1016/j.jenvman.2004.11.024

Chapter 8

HYDROLOGIC DATA ASSIMILATION

Paul R. Houser[1,2], Gabriëlle J.M. De Lannoy[3,4] and Jeffrey P. Walker[5]

[1]George Mason Univ., Fairfax, VA, USA

[2]Bureau of Reclamation, Washington, DC, USA

[3]Ghent Univ., Ghent, Belgium

[4]NASA Goddard Space Flight Center, Greenbelt, MD, USA

[5]Monash University, Melbourne, Australia

INTRODUCTION

Information about hydrologic conditions is of critical importance to real-world applications such as agricultural production, water resource management, flood prediction, water supply, weather and climate forecasting, and environmental preservation. Improved hydrologic condition estimates are useful for agriculture, ecology, civil engineering, water resources management, rainfall-runoff prediction, atmospheric process studies, climate and weather/climate prediction, and disaster management (Houser et al. 2004).

While ground-based observational networks are improving, the only practical way to observe the hydrologic cycle on continental to global scales is via satellites. Remote sensing can make spatially comprehensive measurements of various components of the hydrologic system, but it cannot provide information on the entire system (e.g. evaporation), and the observations represent only an instant in time. Hydrologic process models may be used to predict the temporal and spatial hydrologic variations, but these predictions are often poor, due to model initialization, parameter and forcing, and physics errors. Therefore, an attractive prospect is to combine the strengths of hydrologic models and observations (and minimize the weaknesses) to provide a superior hydrologic state estimate. This is the goal of hydrologic data assimilation.

Data Assimilation combines observations into a dynamical model, using the model's equations to provide time continuity and coupling between the estimated fields. Hydrologic data assimilation aims to utilize both our hydrologic process knowledge, as embodied in a hydrologic model, and information that can be gained from observations. Both model predictions and observations are imperfect and we wish to use both synergistically to obtain a more accurate result. Moreover, both contain different kinds of information, that when used together, provide an accuracy level that cannot be obtained individually.

Figure 1 illustrates the hydrologic land surface data assimilation challenge to merge the spatially comprehensive remote sensing observations with the dynamically complete but typically poor predictions of a hydrologic Land Surface Model (LSM) to yield the best possible hydrological system state estimation. In this illustration, the LSM is a component of a General Circulation Model (GCM) or Earth System Model (ESM). Model biases can be mitigated using a complementary calibration and parameterization process. Limited point measurements are often used to calibrate the model(s) and validate the assimilation results (Walker and Houser 2005).

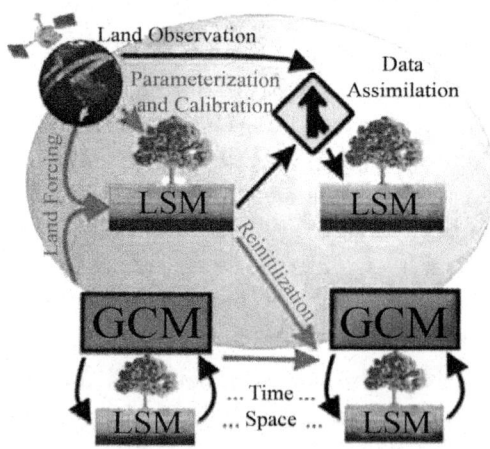

Figure 1. Schematic description of the data assimilation process in a land surface model coupled to a general circulation model.

In this Chapter, we will first provide background on hydrologic observation, modelling and data assimilation. Next we will discuss various hydrologic data assimilation challenges, and finally conclude with several case studies that use hydrologic data assimilation to address disaster management issues.

BACKGROUND: HYDROLOGIC OBSERVATIONS

Remote sensing has transformed our appreciation and modelling of the Earth system over, particularly in the meteorological and oceanographic sciences.

However, historically, remote sensing data have not been widely used in land surface hydrology. This can be ascribed to: (i) a lack of focused hydrologic state (water and energy) remote sensing instruments; (ii) insufficient retrieval algorithms for deriving hydrologic information from remote sensing; (iii) a lack of distributed hydrologic models for incorporating remote sensing information; and (iv) a lack of techniques to objectively improve and constrain hydrologic model predictions using remote sensing. Remote sensing observations have been used in hydrologic models in several ways: (i) to assign parameter input data such as soil and land cover properties; (ii) to assign better atmospheric forcing conditions, such as precipitation, (iii) to set model initial conditions data, such as soil moisture; and (iv) as time-varying land state data, such as snow water content, to constrain model predictions.

Table 1. Characteristics of remotely sensed hydrological observations potentially available within the next decade.

Class	Observation	Ideal Technique	Ideal Time Scale	Ideal Space Scale	Currently available data
Parameters	Land cover/change	optical/IR	daily or changes	1km	AVHRR, MODIS, NPOESS
	Leaf area & greenness	optical/IR	daily or changes	1km	AVHRR, MODIS, NPOESS
	Albedo	optical/IR	daily or changes	1km	MODIS, NPOESS
	Emissivity	optical/IR	daily or changes	1km	MODIS, NPOESS
	Vegetation structure	lidar	daily or changes	100m	ICESAT
	Topography	in-situ survey, radar	changes	1m–1km	GTOPO30, SRTM
Forcings	Precipitation	microwave/IR	hourly	1km	TRMM, GPM, SSMI, GEO-IR, NPOESS
	Wind profile	Radar	hourly	1km	QuickSCAT
	Air humidity & temp	IR, microwave	hourly	1km	TOVS, AIRS, GOES, MODIS, AMSR
	Surface solar radiation	optical/IR	hourly	1km	GOES, MODIS, CERES, ERBS
	Surface LW radiation	IR	hourly	1km	GOES, MODIS, CERES, ERBS
States	Soil moisture	microwave, IR change	daily	1km	SSMI, AMSR, SMOS, NPOESS, TRMM
	Temperature	IR, in-situ	hourly-monthly	1km	IR-GEO, MODIS, AVHRR, TOVS
	Snow cover or SWE	optical, microwave	daily or changes	10m-100m	SSMI, MODIS, AMSR, AVHRR, NPOESS
	Freeze/thaw	radar	daily or changes	10m-100m	Quickscat, IceSAT, CryoSAT
	Ice cover	radar, lidar	daily or changes	10m-100m	IceSAT, GLIMS
	Inundation	optical/microwave	daily or changes	100m	MODIS
	Total water storage	gravity	changes	10km	GRACE
Fluxes	Evapotranspiration	optical/IR, in-situ	hourly	1km	MODIS, GOES
	Streamflow	microwave, laser	hourly	1m-10m	ERS2, TOPEX / POSEIDON, GRDC
	Carbon flux	In-situ	hourly	1km	In-situ
	Solar radiation	optical, IR	hourly	1km	MODIS, GOES, CERES, ERBS
	Longwave radiation	optical, IR	hourly	1km	MODIS, GOES
	Sensible heat flux	IR	hourly	1km	MODIS, ASTER, GOES

The historic lack of remotely sensed hydrological missions and observations has been the result of an historical emphasis on meteorological and oceanographic operations and applications, due to the large scientific and mission communities that drive those fields. However, significant progress has been made over the past decade on defining hydrologically-relevant remote sensing observations through focused ground and airborne field studies. Gradually, remotely-sensed hydrological data are becoming available; land surface skin temperature and snow cover data have been available for many years, and satellite precipitation data are becoming available at increasing space and time resolutions. In addition, land cover/use maps, vegetation parameters (photosynthesis, structure, etc.), and snow observations of increasing sophistication are becoming available from a number of sensors. Novel observations such as saturated fraction and soil moisture changes, evapotranspiration, water level and velocity (i.e., runoff), and changes in total terrestrial water storage are also being developed. Furthermore, near-surface soil moisture, a parameter shown to play a critical role in weather, climate, agriculture, flood, and drought processes, is currently available from non-ideal sensor configuration observations. Moreover, two missions targeted at measuring near-surface soil moisture with ideal sensor configuration are expected before the end of the decade (SMOS and SMAP; see Table 1).

BACKGROUND: HYDROLOGIC MODELLING

Advances in understanding of soil-water dynamics, plant physiology, micrometeorology and the hydrology that control biosphere-atmosphere interactions have spurred the development of hydrologic Land Surface Models (LSMs), whose aim is to represent simply, yet realistically, the transfer of mass, energy and momentum between a vegetated surface and the atmosphere (Sellers et al., 1986). LSM predictions are regular in time and space, but these predictions are influenced by errors in model structure, input variables, parameters and inadequate treatment of sub-grid scale spatial variability. These models are built upon the analysis of signals entering and leaving the system; they predict relationships between physical system variables as a solution of mathematical structures, like simple algebraic equations or differential equations. Hydrologic processes are part of the total of global processes controlling the earth, which are typically represented in global general circulation models (GCMs). The major state variables of these models include the water content and temperature of soil moisture, snow and vegetation. Changes in these state variables account for fluxes, e.g., evapotranspiration or runoff. Recently, coupling of hydrological models with vegetation models has received some attention, to serve more specific ecological, biochemical or agricultural purposes.

Most LSMs used in GCMs view the soil column as the fundamental hydrological unit, ignoring the role of topography on spatially variable processes (Stieglitz et al. 1997) to limit the complexity and computations for these coupled models. Increasingly, LSMs are being built with a higher degree of complexity in order to better represent hydrologic atmosphere interactions within GCMs or to meet the need for local state and process knowledge for use in conservation or agricultural management. This includes the treatment of more biological processes, the representation of subgrid heterogeneity and the development of spatially distributed or gridded models. Improved process representation should result in parameters that are easier to measure or estimate. However, more complex process representations results in more parameters to be estimated, and may lead to over parameterized given the data available for parameter calibration.

Model calibration relies on observed data and can be defined as a specific type of data assimilation, as its goal is to minimize model bias using observations. For large scale hydrologic modelling, full calibration is nearly impossible. Some examples of widely used LSMs are the NCAR Community Land Model (CLM), the Princeton/U. Washington Variable Infiltration Capacity Model (VIC), and the NOAA-Noah Model.

BACKGROUND: HYDROLOGIC DATA ASSIMILATION

Charney et al. (1969) first suggested combining current and past data in an explicit dynamical model, using the model's prognostic equations to provide time continuity and dynamic coupling amongst the fields. This concept has evolved into a family of techniques known as data assimilation. In essence, hydrologic data assimilation aims to utilize both our hydrological process knowledge as embodied in a hydrologic model, and information that can be gained from observations. Both model predictions and observations are imperfect and we wish to use both synergistically to obtain a more accurate result. Moreover, both contain different kinds of information, that when used together, provide an accuracy level that cannot be obtained when used separately.

For example, a hydrological model provides spatial and temporal near-surface and root zone soil moisture information at the model resolution, including error estimates. On the other hand, remote sensing observations contain near-surface soil moisture information at an instant in time, but do not give the temporal variation or the root zone moisture content. While the remote sensing observations can be used as initialization input for models or as independent evaluation, providing we use a hydrological model that has been adapted to use remote sensing data as input, we can use the hydrological model predictions and remote sensing observations together to keep the simulation on track through data assimilation (Kostov and Jackson 1993). Moreover, large errors in near-surface soil moisture content prediction are unavoidable because of its highly dynamic nature. Thus, when measured soil moisture data are available, their use to constrain the simulated data should improve the soil moisture profile estimate, provided that an update in the upper layer is well propagated to deeper layers.

Data assimilation techniques were established by meteorologists (Daley 1991) and have been used very successfully to improve operational weather forecasts. Data assimilation has also been successfully used in oceanography (Bennett 1992) for improving ocean dynamics prediction. However, hydrological data assimilation has a smaller number of case studies demonstrating its utility and has very distinct features compared to atmospheric or oceanographic assimilation. Hydrological data assimilation development has been accelerated by building on knowledge derived from the meteorological and oceanographic data assimilation, with significant recent advancement and increased interdisciplinary interaction.

Hydrologic data assimilation progress has been primarily limited by a lack of suitable largedomain observations. With the introduction of new satellite sensors and technical advances, hydrologic data assimilation research directions are changing (Margulis et al. 2006). Walker et al. (2003) gave a brief history of hydrological data assimilation, focusing on the use and availability of remote sensing data, and stated that this research field is still in its "infancy". Walker and Houser (2005) gave an overview of hydrological data assimilation, discussing different data assimilation methods and several case studies in hydrology. van Loon and Troch (2001) gave a review of hydrological data assimilation applications and added a discussion on the challenges facing future hydrological applications. McLaughlin (1995) reviewed some developments in hydrological data assimilation and McLaughlin (2002) transferred the options of interpolation, smoothing and filtering for state estimation from the engineering to hydrological sciences.

Soil moisture and soil temperature have been the most studied variables for hydrologic model estimation, because of their well-known impact on weather forecasts (Zhang and Frederiksen 2003; Koster et al. 2004) and climate predictions (Dirmeyer 2000). Besides these variables, also snow and vegetation properties have received attention. Hydrologic state variables are highly variable in all three space dimensions, so a complete and detailed assessment of these variables is a difficult task. Therefore, most studies have focused on data assimilation in one or two dimensions (e.g. soil moisture profiles or single layer fields) and/or relatively simple models.

Data assimilation was meant for state estimation, but in the broadest sense, data assimilation refers to any use of observational information to improve a model (WMO 1992). Basically, there are four methods for "model updating", as follows:

- **Input**: corrects model input forcing errors or replaces model-based forcing with observations, thereby improving the model's predictions;
- **State**: corrects the state or storages of the model so that it comes closer to the observations (state estimation, data assimilation in the narrow sense);
- **Parameter**: corrects or replaces model parameters with observational information (parameter estimation, calibration);
- **Error correction**: correct the model predictions or state variables by an observed timeintegrated error term in order to reduce systematic model bias (e.g. bias correction).

The data assimilation challenge is: given a (noisy) model of the system dynamics, find the best estimates of system states from (noisy) observations. Most current approaches to this problem are derived from either the direct observer (i.e., sequential filter) or dynamic observer (i.e., variational through time) techniques. Figure 2 illustrates schematically the key differences between these two approaches to data assimilation. To help the reader through the large amount of jargon typically associated with data assimilation, a list of terminology has been provided (Table 2).

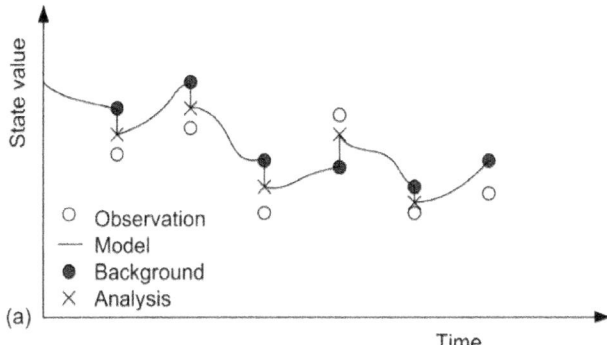

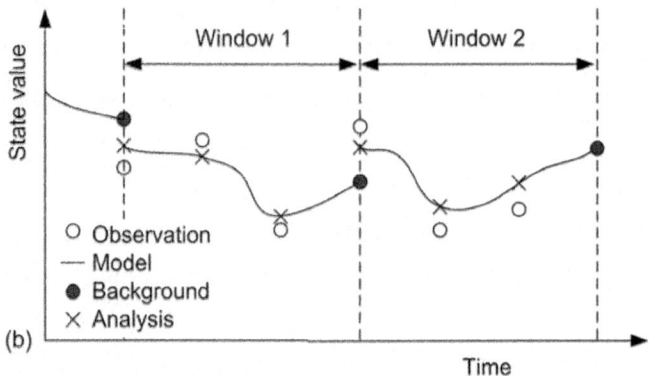

Figure 2. Schematic of the (a) direct observer and (b) dynamic observer assimilation approaches.

Table 2. Commonly used data assimilation terminology

State	condition of a physical system, e.g. soil moisture
State error	deviation of the estimated state from the truth
Prognostic	a model state required to propagate the model forward in time
Diagnostic	a model state/flux diagnosed from the prognostic states – not required to propagate the model
Observation	measurement of a model diagnostic or prognostic
Covariance matrix	describes the uncertainty in terms of standard deviations & correlations
Prediction	model estimate of states
Update	correction to a model prediction using observations
Background	forecast, prediction or state estimate prior to an update
Analysis	state estimate after an update
Innovation	observation-prediction, a priori residual
Gain matrix	correction factor applied to the innovation
Tangent linear model	linearized (using Taylor's series expansion) version of a non-linear model
Adjoint	operator allowing the model to be run backwards in time

Data assimilation has significant benefits beyond the improved state estimates, as follows (adapted from Rood et al. 1994).

- **Organizes**: By interpolating information from observation space to model space, the observations are organized and given dynamical consistency with the model equations, thereby enhancing their usefulness;

- **Supplements**: By constraining the model's physical equations with observations, unobserved quantities can be better estimated, providing a more complete understanding of the true hydrological system;

- **Complements**: By propagating information using observed spatial and temporal correlations, or the model's physical relationships, areas of sparse observations can be better estimated;

- **Quality control**: By comparing observations with previous forecasts, spurious observations can be identified and eliminated. By performing this comparison over time, it is possible to calibrate observing systems and identify biases or changes in observation system performance;

- **Hydrological model improvement**: By continuously confronting the model with real observations, model weaknesses and systematic errors can be identified and corrected.

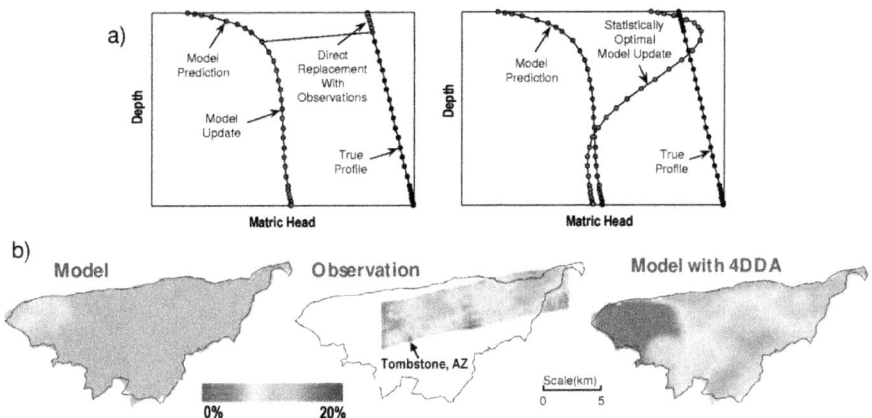

Figure 3. Example of how data assimilation supplements data and complements observations: a) Numerical experiment results demonstrating how near-surface soil moisture measurements are used to retrieve the unobserved root zone soil moisture state using (left panel) direct insertion and (right panel) a statistical assimilation approach (Walker et al. 2001a); b) Six Push Broom Microwave Radiometer (PBMR) images gathered over the USDAARS Walnut Gulch Experimental Watershed in southeast Arizona were assimilated into the TOPLATS hydrological model using several alternative assimilation procedures (Houser et al. 1998). The observations were found to contain horizontal correlations with length scales of several tens of km, thus allowing soil moisture information to be advected beyond the area of the observations.

HYDROLOGIC DATA ASSIMILATION TECHNIQUES

Direct insertion. One of the earliest and most simplistic approaches to data assimilation is direct insertion. As the name suggests, the forecast model states are directly replaced with the observations. This approach makes the explicit assumption that the model is wrong (has no useful information) and that the observations are right, which both disregards important information provided by the model and preserves observational errors. The risk of this approach is that unbalanced state estimates may result, which causes model shocks: the model will attempt to restore the dynamic balance that would have existed without insertion. A further key disadvantage of this approach is that model physics are solely relied upon to propagate the information to unobserved parts of the system (Houser et al. 1998; Walker et al. 2001a).

Statistical Correction

A derivative of the direct insertion approach is the statistical correction approach, which adjusts the mean and variance of the model states to match those of the observations. This approach assumes the model pattern is correct but contains a non-uniform bias. First, the predicted observations are scaled by the ratio of observational field standard deviation to predicted field standard deviation. Second, the scaled predicted observational field is given a block shift by the difference between the means of the predicted observational field and the observational field (Houser et al. 1998). This approach also relies upon the model physics to propagate the information to unobserved parts of the system.

Successive Correction

The successive corrections method (SCM) was developed by Bergthorsson and Döös (1955) and Cressman (1959), and is also known as observation nudging. The scheme begins with an a priori state estimate (background field) for an individual (scalar) variable, which is successively adjusted by nearby observations in a series of scans (iterations, n) through the data. The analysis at time step k is found by passing through a sequence of updates.

The advantage of this method lies in its simplicity. However, in case of observational error or different sources (and accuracies) of observations, this scheme is not a good option for assimilation, since information on the observational accuracy is not accounted for. Mostly, this approach assumes that the observations are more accurate than model forecasts, with the observations fitted as closely as is consistent. Furthermore, the radii of influence are user-defined and should be determined by trial and error or more sophisticated methods that reduce the advantage of its simplicity. The weighting functions are empirically chosen and are not derived based on physical or statistical properties. Obviously, this method is not effective in data sparse regions. Some practical examples are discussed by Bratseth (1986) and Daley (1991).

Analysis Correction

This is a modification to the successive correction approach that is applied consecutively to each observation s from 1 to sf as in Lorenc et al. (1991). In practice, the observation update is mostly neglected and further assumptions make the update equation equivalent to that for optimal interpolation (Nichols 1991).

Nudging

Nudging or Newtonian relaxation consists of adding a term to the prognostic model equations that causes the solution to be gradually relaxed towards the observations. Nudging is very similar to the successive corrections technique and only differs in the fact that through the numerical model the time dimension is included. Two distinct approaches have been developed (Stauffer and Seaman 1990). In analysis nudging, the nudging term for a given variable is proportional to the difference between the model simulation at a given grid point and an "analysis" of observations (i.e., processed observations) calculated at the corresponding grid point. For observation nudging, the difference between the model simulation and the observed state is calculated at the observation locations.

Optimal Interpolation

The optimal interpolation (OI) approach, sometimes referred to as statistical interpolation, is a minimum variance method that is closely related to kriging. OI approximates the "optimal" solution often with a "fixed" structure for all time steps, given by prescribed variances and a correlation function determined only by distance (Lorenc 1981). Sometimes, the variances are allowed to evolve in time, while keeping the correlation structure time-invariant.

3-D Variational

This approach directly solves the iterative minimization problem given (Parrish and Derber 1992). The same approximation for the background covariance matrix as in the optimal interpolation approach is typically used.

Kalman Filter

The optimal analysis state estimate for linear or linearized systems (Kalman or Extended Kalman filter, EKF) can be found through a linear update equation with a Kalman gain that aims at minimizing the analysis error (co) variance of the analysis state estimate (Kalman 1960). The essential feature which distinguishes the family of Kalman filter approaches from more static techniques, like optimal interpolation, is the dynamic updating of the forecast (background) error covariance through time. In the traditional Kalman filter (KF) approach this is achieved by application of standard error propagation theory, using a (tangent) linear model. (The only difference between the

Kalman filter and the Extended Kalman filter is that the forecast model is linearized using a Taylor series expansion in the latter; the same forecast and update equations are used for each approach.)

A further approach to estimating the state covariance matrix is the Ensemble Kalman filter (EnKF). As the name suggests, the covariances are calculated from an ensemble of state forecasts using the Monte Carlo approach rather than a single discrete forecast of covariances (Turner et al. 2007).

Reichle et al. (2002b) applied the Ensemble Kalman filter to the soil moisture estimation problem and found it to perform as well as the numerical Jacobian approximation approach to the Extended Kalman filter, with the distinct advantage that the error covariance propagation is better behaved in the presence of large model non-linearities. This was the case even when using only the same number of ensembles as required by the numerical approach to the Extended Kalman filter.

4D-Var

In its pure form, the 4-D (3-D in space, 1-D in time) "variational" (otherwise known as Gauss-Markov) dynamic observer assimilation methods use an adjoint to efficiently compute the derivatives of the objective function with respect to each of the initial state vector values. Solution to the variational problem is then achieved by minimization and iteration. In practical applications the number of iterations is usually constrained to a small number. While "adjoint compilers" are available for automatic conversion of the non-linear forecast model into a tangent linear model, application of these is not straightforward. It is best to derive the adjoint at the same time as the model is developed.

Given a model integration with finite time interval, and assuming a perfect model, 4D-Var and the Kalman filter yield the same result at the end of the assimilation time interval. Inside the time interval, 4D-Var is more optimal, because it uses all observations at once (before and after the time step of analysis), i.e., it is a smoother. A disadvantage of sequential methods is the discontinuity in the corrections, which causes model shocks. Through variational methods, there is a larger potential for dynamically based balanced analyses, which will always be situated within the model climatology. Operational 4D-Var assumes a perfect model: no model error can be included. With the inclusion of model error, coupled equations are to be solved for minimization. Through Kalman filtering it is in general simpler to account for model error.

Both the Kalman filter and 3D/4D-Var rely on the validity of the linearity assumption. Adjoints depend on this assumption and incremental 4D-Var is even more sensitive to linearity. Uncertainty estimates via the Hessian are critically dependent on a valid linearization. Furthermore, with variational assimilation it is more difficult to obtain an estimate of the quality of the analysis or of the state's uncertainty after updating.

ASSIMILATION OF HYDROLOGIC OBSERVATIONS

Estimation of the hydrologic state has mainly been focused on soil moisture, snow water content, and temperature. The observations used to infer state information range from direct field measurements of these quantities to more indirectly related measurements like radiances or backscatter values in remote sensing products. A few studies have also tried to assimilate state-dependent diagnostic fluxes, like discharge or remotely sensed heat fluxes. The success of assimilation of observations which are indirectly related to the state is largely dependent on a good characterization of the observation operator. This section presents examples of research in hydrologic data assimilation, but is not intended to be a comprehensive review.

Truly optimal data assimilation techniques require flawless model and observation error characterization. Therefore, recent studies have focused on the first and second order error characterization in hydrologic modelling. Typically, either model predictions or observations are biased. Studies by Reichle and Koster (2004), Bosilovich et al. (2007) and De Lannoy et al. (2007a, b) scratch the surface of how to deal with these hydrologic modelling biases. The second order error characterization is of major importance to optimize the analysis result and for the propagation of information through the system. Tuning of the error covariance matrices has, therefore, gained attention with the exploration of adaptive filters in hydrologic modelling (Reichle et al. 2008; De Lannoy et al. 2009).

Furthermore, it is important to understand that hydrologic data assimilation applications are dealing with non-closure or imbalance problems, caused by external data assimilation for state estimation. In a first attempt to attack this problem, Pan and Wood (2006) developed a constrained Ensemble Kalman filter which optimally redistributes any imbalance after conventional filtering. They applied this technique over a 75,000 km2 domain in the US, using the terrestrial water balance as constraint.

CASE STUDIES

Significant advances in hydrological data assimilation have been made over the past decade from which we have selected a few case studies to demonstrate the utility of hydrological data assimilation in hazard prediction and mitigation.

Case study 1: Assimilation of Water Level Observations

By providing predictions of flood hazard and risk over increasing lead-times, flood inundation models play a central role in advanced hydro-meteorological forecasting systems. As the cost of damage caused by flooding is highly dependent on the warning time given before a storm event, the reduction of its predictive uncertainties has received a great deal of attention by researchers in recent years (e.g., Montanari et al., 2009, Biancamaria et al., 2010). The predictive uncertainty originates from several causes interacting between each other, namely input uncertainty (i.e. inflows), model structure and parameter uncertainty. The predictive uncertainty can be reduced through a periodical updating of computed water surface lines by taking advantage of water level measurements. However, ground based data are spatially rather limited, numbers of hydrometric stations are in decline at a global scale and major parts of the world still remain largely ungauged to this date. Recent developments in remotely sensing-based measurement techniques potentially help overcoming data scarcity. For instance, the technique of water stage retrievals from satellite measurements with centimeter-scale accuracy (e.g. Alsdorf et al., 2007) can be seen as a promising alternative to hydrometric station data.

Matgen et al. (2010) demonstrated that the real-time assimilation of remote sensingderived water elevation into 1D hydraulic models via a Particle Filter enables the correction of water depth from a corrupted hydraulic model. In their synthetic experiments they found that significant model improvements could be achieved with observation error standard deviations up to 5 m. Another interesting result from their synthetic experiments is the realization that it is crucial to adjust the fluxes at the upstream boundaries of the model in order to significantly and persistently improve the hydraulic model. In river hydraulics, the process time scale is relatively short, so that stock updates have a limited lifetime. As in Andreadis et al. (2007), the research of Matgen et al. (2010) has clearly demonstrated that because of the dominating effect of the upstream boundary condition merely updating the state variable of the model (water level and hence water storage), only improves the model forecast over a very short time horizon. Model predictions rapidly degrade after updating if the forcing data are not consistent with observed water levels. Updating the uncertain upstream boundary condition leads to more persistent model improvement.

Giustarini et al. (2011) recently tested the methodology with real event data using water level data obtained from ERS-2 SAR and ENVISAT ASAR during the January 2003 flood of the Alzette River (Grand-Duchy of Luxembourg). The retrieval of water elevation data from SAR is based on three steps (see Hostache et al., 2009 for a detailed description of the method). First, the flood extension limits with their respective geolocation uncertainty are derived from a SAR image using a radiometric thresholding-based procedure. Next, the resulting uncertain flood extent limits are superimposed on a digital elevation model (DEM) in order to estimate local water levels. The method takes into account the uncertainty stemming from the underlying DEM. The water level information is obtained as model cross-section specific intervals of the possible local water level. In the last step, the intervals of water levels are hydraulically constrained in order to reduce the estimation uncertainty (see Figure 4). The Particle Filter-based assimilation scheme consists in having a single particle with water levels at all cross sections as state vector. Hence, the likelihood that is computed for each particle is derived from its ability to correctly predict water levels along the entire river reach. In order to overcome the problem of non-persistent model improvements, the forcing of the hydraulic model is updated as well, using information on model error that is obtained during the analysis cycle.

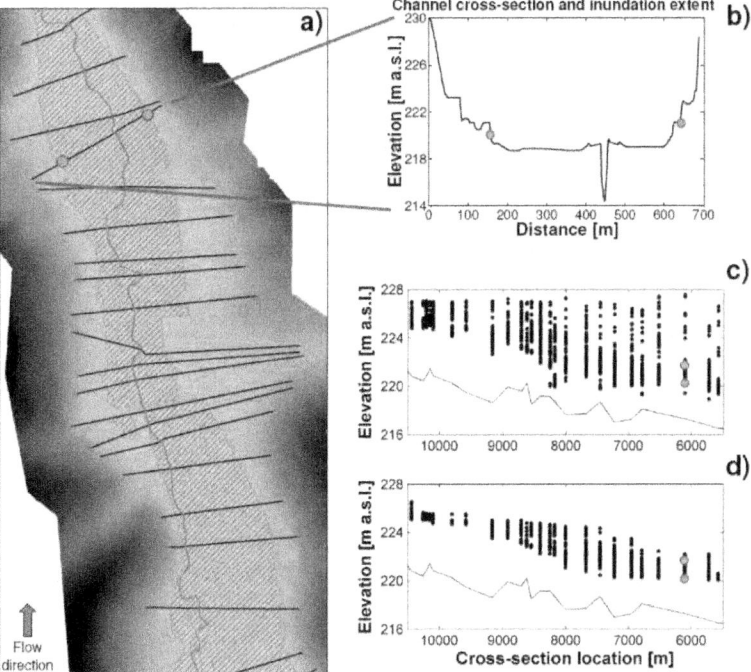

Figure 4. Diagram showing an example of: (a) flood extent derived from a satellite image superimposed on the DEM and the river cross-section location, (b) illustration of water level values extracted for a given cross-section and c) the remote sensing-derived water elevation along a portion of stream (c) before and (d) after applying the hydraulic coherence constrain.

The approach works well if inflows are the main source of error in hydraulic modelling. However, when model behavior is nonuniform across the model domain, it is preferred to have the Particle Filter assign a separate particle set to each cross-section. Giustarini et al. (2011) further conclude that the analysis step is of major importance for carrying out an efficient inflow correction over many time steps as errors in the analysis will propagate through the inflow correction model, thereby potentially degrading the skill of the forecasts. The data assimilation experiments show the potential of remote sensing-derived water level data for persistently improving model predictions over many time steps (see Figure 5).

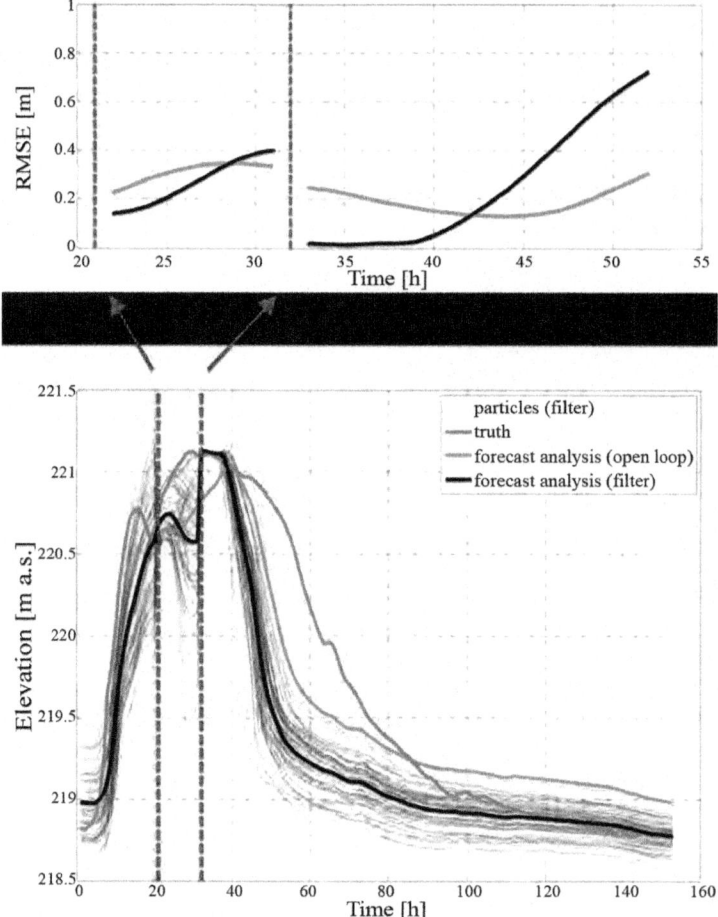

Figure 5. Stage hydrographs at one river cross-section before and after assimilating remote sensing-derived water level intervals from two SAR images into the 1D hydraulic model (bottom panel). The forecasting performance is evaluated with the RMSE

evolution in time (top panel). The cyan line represents the RMSE before assimilation and the black line displays the RMSE after assimilation.

Case study 2: Assimilation of Snow Water Equivalent and Snow Cover Fraction

Snowmelt runoff is of major importance to summer water supplies, and plays a considerable role in mid-latitude flood events. Snow alters the interface between the atmosphere and the land surface through its higher albedo and lower roughness compared to snow-free conditions, and by thermally insulating the soil from the atmosphere. Consequently, the presence of snow strongly affects the land surface water and energy balance, weather and climate. Moreover, snow has a high spatial and temporal variability which is very sensitive to global change.

Numerical simulation of snow processes is far from perfect (Slater et al., 2001), therefore snow data assimilation could provide a more accurate estimate of snow conditions. Satellitebased snow cover fraction (SCF) observations are available using visible and near-infrared measurements from sensors like the Moderate Resolution Imaging Spectroradiometer (MODIS, 2000 - present). While accurate, these observations have limitations (Dong and Peters-Lidard, 2010), such as the inability to see through clouds. Additionally, SCF observations only provide a partial estimate of the snow state, namely snow cover; in contrast, hydrologic modeling uses snow water equivalent (SWE, snow mass), so snow depletion curves are used to imperfectly translate SCF to SWE. These SCF issues can be overcome using SWE observations derived from passive microwave observations such as the Scanning Multichannel Microwave Radiometer (SMMR, 1978 – 1987), Special Sensor Microwave Imager (SSM/I, 1987-present) and Advanced Microwave Scanning Radiometer for the Earth Observing System (AMSR-E, 2002-present). These sensors do not suffer from cloud obscuration and allow SWE estimation by relating the microwave brightness temperature to snow parameters, but typically have a much coarser resolution and low precision (Dong et al., 2005).

Therefore, an improved snow analyses can be expected by combining the strengths of different snow observations and models. De Lannoy et al. (2011) examined the possibilities and limitations of assimilating both fine-scale MODIS SCF and coarse-scale AMSR-E SWE retrievals into the Noah LSM using an EnKF. Eight years (2002–2010) of remotely sensed AMSR-E snow water equivalent (SWE) retrievals and MODIS snow cover fraction (SCF) observations were assimilated into the Noah LSM over a domain in Northern Colorado using a multi-scale ensemble Kalman filter (EnKF), combined with a rule-based update. De Lannoy et al., 2011 discuss several experiments: (a) ensemble open loop without assimilation (EnsOL); (b) assimilation of coarse-scale AMSR-E SWE observations (SWE DA); (c) assimilation of fine-scale MODIS SCF observations (SCF DA), which involves a mapping from SCF to SWE; and (d) joint, multi-scale assimilation of AMSR-E SWE and MODIS SCF observations (SWE & SCF DA).

Figure 6 illustrates the spatial patterns of the satellite observations, the EnsOL estimates, and the assimilation estimates (without scaling) for a few representative days during the winter of 2009-2010. For this winter, the model and satellite observations have a similar SWE magnitude and no explicit bias-correction is needed to interpret the spatial patterns. The 3D filter performs a downscaling of the coarse AMSR-E SWE observations and shows a realistic fine-scale variability driven by the land surface model integration (De Lannoy et al., 2010). For example, high elevations maintain SWE values well above the observed AMSR-E SWE, which would not be the case if the AMSR-E pixels were a priori partitioned and assimilated with a 1D filter. Furthermore, areas without observations (swath effects) are updated through spatial correlations in the forecast errors. The 1D SCF filter imposes the fine-scale MODIS-observed variability, and locations without fine-scale observations (due to clouds) are not updated. The combined SWE and SCF assimilation shows features of both the SWE and SCF assimilation integrations. Assimilation and downscaling of coarse-scale AMSR-E SWE as well as MODIS SCF assimilation maintain realistic spatial patterns in the SWE analyses.

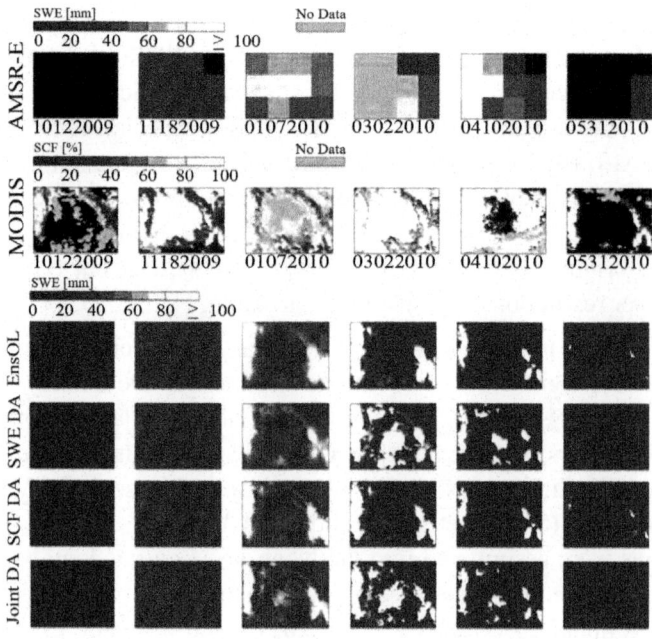

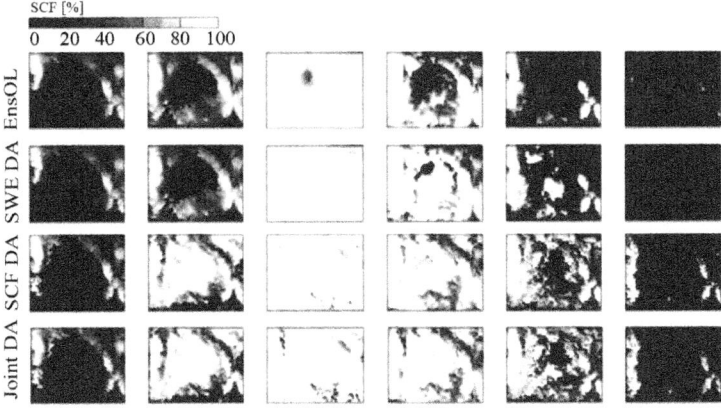

Figure 6. SWE (at 8:00 UTC) and SCF (at 17:00 UTC) fields at 5 days in the winter of 2009-2010. The top 2 series show the assimilated observations, the other plots show SWE and SCF for the EnsOL forecast and 3 different data assimilation (DA) analyses. AMSR-E data are missing due to the swath effect and MODIS data are missing because of cloud contamination.

Case study 3: Assimilation of Soil Moisture Observations

Soil moisture is an important initialization variable for large-scale weather forecasts and climate predictions. Smaller-scale soil moisture conditions have a great impact on agriculture, ecology and hydrology. The occurrence of droughts and flooding has a major impact on human lives and monitoring the soil moisture helps to prevent or mitigate disasters. Regional to global soil moisture data rely mostly on land surface model simulations forced by meteorological data or on satellite-based microwave observations. However, these satellite observations have coarse spatial resolution, only sense the top few centimetres of the soil and are only available for a specific area when the satellite happens to pass over. Through assimilation of these intermittent surface observations into a model, continuous and consistent soil moisture profile estimates could be obtained.

Liu et al. (2011) illustrated how land surface simulations can be improved by either improving the precipitation, or by assimilating surface soil moisture retrievals from the Advanced Microwave Scanning Radiometer (AMSR-E) with a 1-D ensemble Kalman filter into the NASA Catchment land surface model. The assimilated soil moisture products were either the operational NASA Level-2B AMSR-E "AE-Land" product (archived by NSIDC), or the Land Parameter Retrieval Model C-band LPRM-C product. The forcings were based on the atmospheric forcing fields from Modern Era Retrospectiveanalysis for Research and Applications (MERRA), but the precipitation is corrected with large-scale, gauge- and satellite-based precipitation observations from different datasets (CMAP, GPCP, and CPC). The soil moisture skill was defined as the anomaly time series correlation coefficient R of the model or assimilation results against in situ observations in the continental United States at 44 single-profile sites within the Soil Climate Analysis Network (SCAN). Figure 7 shows that the precipitation corrections and assimilation of satellite soil moisture retrievals contribute comparable and largely independent amounts of information to the assimilation results. Furthermore, it should be stressed that the satellite observations are only available for the surface soil layer and assimilation of these surface data clearly helps to improve the soil moisture estimates in the root zone.

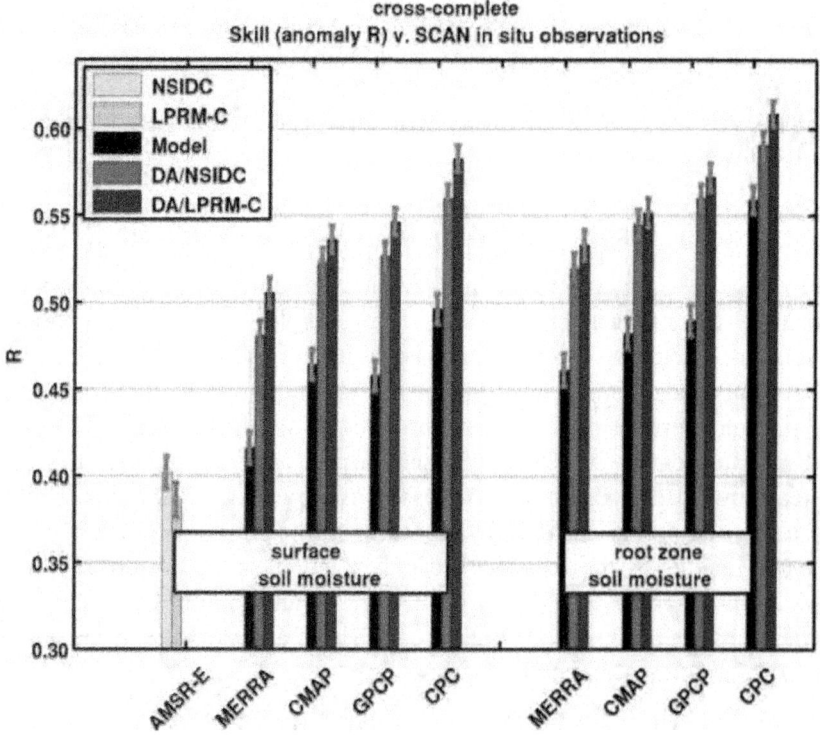

Figure 7. Average time series correlation coefficient R with SCAN in situ surface and root zone soil moisture anomalies for estimates from two AMSR-E retrievals (NSIDC and LPRM-C), the Catchment model forced with four different precipitation datasets (MERRA, CMAP, GPCP, and CPC), and the corresponding data assimilation integrations (DA/NSIDC and DA/LPRM-C). Average is over 44 SCAN sites for surface soil moisture and over 42 sites for root zone soil moisture. Error bars indicate approximate 95% confidence intervals.

The above example assimilated the coarse-scale AMSR-E data with a 1-D filter and focused on improving the temporal characteristics of the assimilation results for large scale applications. In another study by Sahoo et al. (2011), coarse-scale AMSR-E observations were assimilated with more focus on the fine-scale spatial variability by applying a 3-D spatial filter (Reichle and Koster, 2003, De Lannoy et al., 2010) to downscale the observations to the fine-scale model resolution over the Little River Experimental Watershed in Georgia. A correct assessment of the soil moisture pattern could largely impact flood predictions and may be crucial in the effective mitigation of droughts. Furthermore, as numerous previous studies (e.g., Walker et al. (2001b), De Lannoy et al. (2006), Liu et al. (2011)), it was reconfirmed that the assimilation of surface observations impact the deeper layer soil moisture and other water balance variables, but Sahoo et al. (2011) also illustrated that assimilation of surface observations helps the model to spin up faster to its balanced state, both in the surface and deeper layers. This is shown by the gray arrow in Figure 8, which indicates the time difference in model spin up without and with data assimilation. Data assimilation thus better prepares and balances the land surface models to provide improved short-term soil moisture forecasts.

Case study 4: Soil Moisture Assimilation and NWP

Soil moisture can influence the development of the low-level atmosphere, by controlling the partition of incoming radiation into latent and sensible heat fluxes. In Numerical Weather Prediction (NWP) models, errors in the model forecasts (particularly from precipitation) tend to accumulate in the model soil moisture states, causing the modelled soil moisture to gradually drift away from reality. At many NWP centers this is prevented by applying simple nudging or Optimal Interpolation-based assimilation schemes that correct the model soil moisture to reduce errors in forecasts of low-level relative humidity and atmospheric temperature, based on screen-level (1.5 - 2m) observations from automatic weather stations. While this approach can effectively reduce low-level atmospheric forecast errors (of greatest concern to NWP) this is often achieved by degrading the model soil moisture, since it is 'corrected' to compensate for screen-level errors unrelated to soil moisture, for example due to inaccuracies in the land surface flux parameterisations or the radiation physics (Drusch et al, 2007).

Ultimately, inaccurate soil moisture in an NWP model will lead to inaccurate atmospheric forecasts. Additionally, if accurate soil moisture states could be obtained from NWP models, these would be valuable for other operational applications, such as hydrological modelling, flood forecasting, and drought monitoring. A promising solution to improving the accuracy of the soil moisture in NWP models is to make use of novel remotely sensed observations of near-surface soil moisture, such as those available from AMSR-E. Within this context, a study by Draper et al. (2011) presents an Extended Kalman Filter (EKF) capable of assimilating both screen-level observations and remotely sensed near-surface soil moisture observations into an NWP model. This EKF, based on the Simplified EKF of Mahfouf et al (2009), was specifically designed to be computationally affordable within an operational NWP model, however the experiments presented here were conducted using an offline land surface model (with no feedback to the atmospheric model).

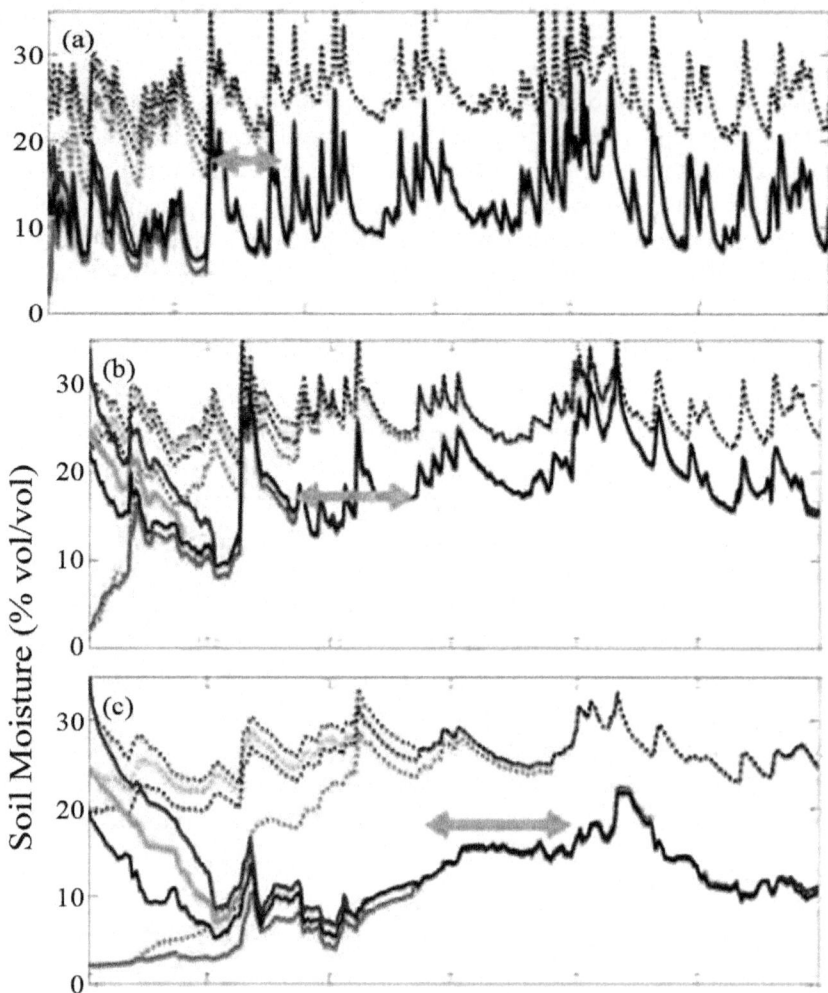

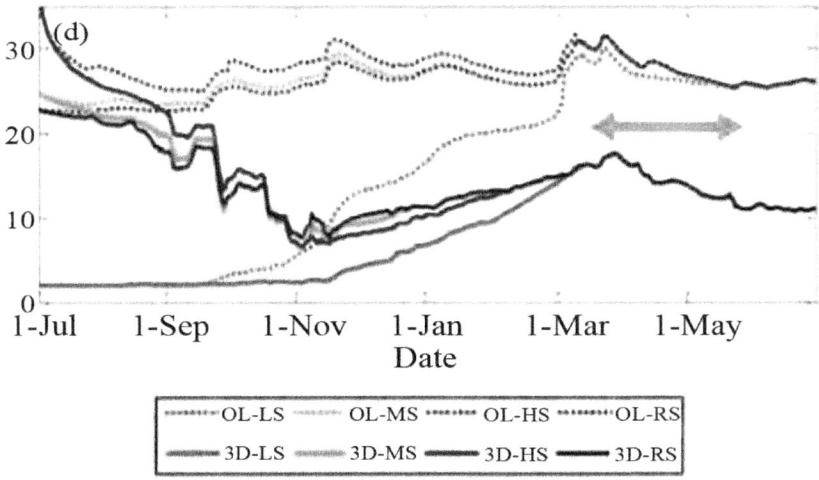

Figure 8. Sensitivity results of the EnKF 3-D algorithm to the model initialization conditions and model spin-up for the Soil Moisture (a) Layer 1, (b) Layer 2, (c) Layer 3 and (d) Layer 4. The results are spatially averaged over 16 in-situ locations. OL (dashed) stands for the Open Loop model simulation without assimilation, started from different initial soil moisture wetness values. 3D (solid line) refers to assimilation results with L (low), M (moderate) and H (high) initial soil moisture. The simulations start in the summer (July 1, 2002 = S).

A series of assimilation experiments was conducted to compare the EKF assimilation of AMSR-E derived near-surface soil moisture and screen-level observations into MétéoFrance's NWP model over Europe. Figure 9 demonstrates how assimilating each data set influenced the fit between the subsequent model forecasts and each of the assimilated data sets. When the AMSR-E soil moisture and screen-level observations were assimilated separately, there was no clear consistency between the resulting root-zone soil moisture analyses, and so Figure 9 shows that assimilating one data set did not improve the model fit to the other data set. Hence, for these experiments the screen-level observations could not have been substituted with the AMSR-E data to achieve similar corrections to the low-level atmospheric forecasts, implying that the remotely sensed soil moisture may not be immediately useful for Météo-France's NWP model. However, for the experiments assimilating the screen-level observations the soil moisture innovations were dominated by a diurnal cycle that was not related to the model soil moisture, reinforcing the need to develop the assimilation of alternative data sets.

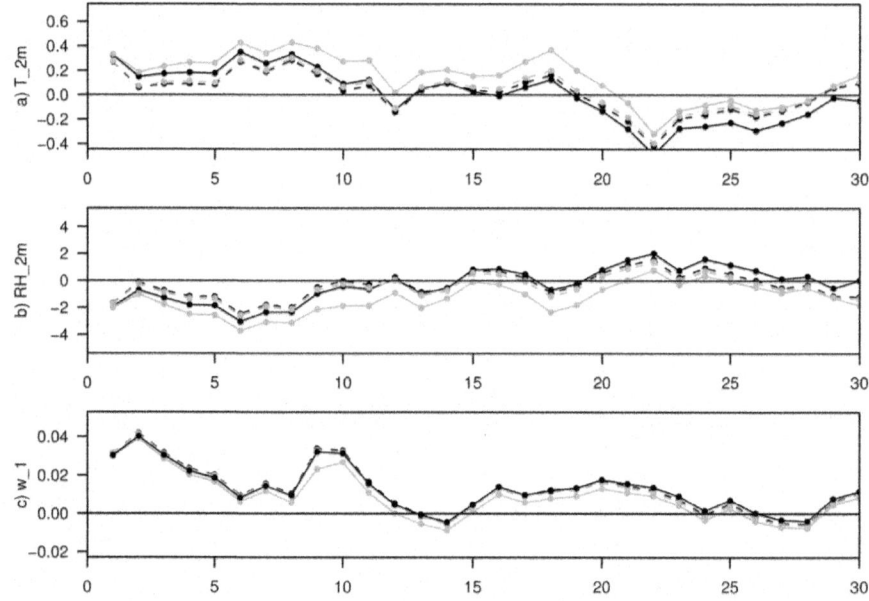

Figure 9. Mean daily observation minus model forecast, averaged over Europe, for each day in July 2006, for a) temperature at 2m above the surface, T_2m (K), b) relative humidity at 2m above the surface, RH_2m (%), and c) surface soil moisture, w_1 (m3m-3), for an open-loop (no assimilation; black, solid), assimilation of screen-level variables (black, dashed), assimilation of AMSR-E soil moisture (grey, solid), and assimilation of both data sets (grey, dashed) experiments.

When the AMSR-E and screen-level observations were assimilated together, the EKF slightly improved the fit between the model forecasts and both observed data sets, although these improvements were very modest, and the root mean square difference over the one month experiment over all of Europe between the model forecasts and the assimilated observations was reduced by less than 5% of the open-loop values, for all assimilated variables. If this result can be substantiated with larger improvements by performing the assimilation in a fully coupled NWP model, this would confirm that assimilating remotely sensed nearsurface soil moisture together with screen-level observations has the potential to improve the realism of the NWP land surface without degrading the low-level atmospheric forecasts.

SUMMARY

Hydrological data assimilation is an objective method to estimate the hydrological system states from irregularly distributed observations. These methods integrate observations into numerical prediction models to develop physically consistent estimates that better describe the hydrological system state than the raw observations alone. This process is extremely valuable for providing initial conditions for hydrological system prediction and/or correcting hydrological system prediction, and for increasing our understanding and improving parametrization of hydrological system behaviour through various diagnostic research studies.

Hydrological data assimilation has still many open areas of research. Development of hydrological data assimilation theory and methods is needed to: (i) better quantify and use model and observational errors; (ii) create model-independent data assimilation algorithms that can account for the typical non-linear nature of hydrological models; (iii) optimize data assimilation computational efficiency for use in large operational hydrological applications; (iv) use forward models to enable the assimilation of remote sensing radiances directly; (v) link model calibration and data assimilation to optimally use available observational information; (vi) create multivariate hydrological assimilation methods to use multiple observations with complementary information; (vii) quantify the potential of data assimilation downscaling; and (viii) create methods to extract the primary information content from observations with redundant or overlaying information. Further, the regular provision of snow, soil moisture, and surface temperature observations with improved knowledge of observational errors in time and space are essential to advance hydrological data assimilation. Hydrological models must also be improved to: (i) provide more "observable" land model states, parameters, and fluxes; (ii) include advanced processes such as river runoff and routing, vegetation and carbon dynamics, and groundwater interaction to enable the assimilation of emerging remote sensing products; (iii) have valid and easily updated adjoints; and (iv) have knowledge of their prediction errors in time and space. The assimilation of additional types of hydrological observations, such as streamflow, vegetation dynamics, evapotranspiration, and groundwater or total water storage must be developed.

As with most current data assimilation efforts, we describe data assimilation procedures that are implemented in uncoupled models. However, it is well known that the highresolution time and space complexity of hydrological phenomena have significant interaction with atmospheric, biogeochemical, and oceanic processes. Scale truncation errors, unrealistic physics formulations, and inadequate coupling between hydrology and the overlying atmosphere can produce feedbacks that can cause serious systematic hydrological errors. Hydrological balances cannot be adequately described by current uncoupled hydrological data systems, because large analysis increments that compensate for errors in coupling processes (e.g. precipitation) result in important non-physical contributions to the energy and water budgets. Improved coupled process models with improved feedback processes, better observations, and comprehensive methods for coupled assimilation are needed to achieve the goal of fully coupled data assimilation systems that should produce the best and most physically consistent estimates of the Earth system.

REFERENCES

1. Alsdorf, D. E., Rodriguez, E., & Lettenmaier, D. P. (2007). Measuring surface water from space, Reviews of Geophysics, 45, doi:10.1029/2006RG000197.

2. Andreadis, K. M., Clark, E. A., Lettenmaier, D. P., & Alsdorf, D. E. (2007). Prospects for river discharge and depth estimation through assimilation of swath-altimetry into a raster-based hydrodynamics model, Geophys. Res. Lett., 34, L10403, doi:10.1029/2007GL029721.

3. Andreadis, K.M. and D.P. Lettenmaier, 2006. Assimilating remotely sensed snow observation into a macroscale hydrology model. Adv. Water Resour., 29, 872–886.

4. Bennett, A.F., 1992. Inverse methods in physical oceanography. Cambridge University Press, 346 pp.

5. Bergthorsson, P. and B. Döös, 1955. Numerical weather map analysis. Tellus, 7, 329–340.

6. Biancamaria, S., Durand, M., Andreadis, K. M., Bates, P. D., Booneg, A., Mognard, N. M., Rodríguez, E., Alsdorf, D. E., Lettenmaier, D. P., & Clark, E. A. (2010). Assimilation of virtual wide swath altimetry to improve Arctic river modelling, Remote Sensing of Environment, doi:10.1016/j.rse.2010.09.008.

7. Bosilovich, M.G., J.D. Radakovich, A.D. Silva, R. Todling and F. Verter, 2007. Skin temperature analysis and bias correction in a coupled land-atmosphere data assimilation system. J. Meteorol. Soc. Jpn., 85A, 205-228.

8. Bratseth, A.M., 1986. Statistical interpolation by means of successive corrections. Tellus, 38A, 439-447.

9. Charney, J.G., M. Halem and R. Jastrow, 1969. Use of incomplete historical data to infer the present state of the atmosphere. J. Atmos. Sci., 26, 1160-1163.

10. Cressman, G.P., 1959. An operational objective analysis system. Mon. Weather Rev., 87, 367–374.

11. Daley, R., 1991. Atmospheric data analysis. Cambridge University Press, 460 pp.

12. De Lannoy, G.J.M., P.R. Houser, N.E.C. Verhoest and V.R.N. Pauwels, 2009. Adaptive soil moisture profile filtering for horizontal information propagation in the independent column-based CLM2.0, J. Hydrometeorol., 10(3), 766-779.

13. De Lannoy, G.J.M., P.R. Houser, V.R.N. Pauwels and N.E.C. Verhoest, 2006. Assessment of model uncertainty for soil moisture through ensemble verification. J. Geophys. Res., 111, D10101.1-18.

14. De Lannoy, G.J.M., P.R. Houser, V.R.N. Pauwels and N.E.C. Verhoest, 2007b. State and bias estimation for soil moisture profiles by an ensemble Kalman filter: effect of assimilation depth and frequency. Water Resour. Res., 43, W06401, doi:10.1029/2006WR005100.

15. De Lannoy, G.J.M., R. H. Reichle, K. R. Arsenault, P. R. Houser, S. Kumar, N. E. C. Verhoest, V.R.N. Pauwels, 2011. Multi-Scale Assimilation of AMSR-E Snow Water Equivalent and MODIS Snow Cover Fraction in Northern Colorado, Water Resour. Res., in review.

16. De Lannoy, G.J.M., R.H. Reichle, P.R. Houser, K. R. Arsenault, V.R.N. Pauwels and N.E.C. Verhoest, 2009. Satellite-scale snow water equivalent assimilation into a highresolution land surface model, J. Hydrometeorol., in press.

17. De Lannoy, G.J.M., R.H. Reichle, P.R. Houser, V.R.N. Pauwels and N.E.C. Verhoest, 2007a. Correcting for forecast bias in soil moisture assimilation with the ensemble Kalman filter. Water Resour. Res., 43, W09410, doi:10.1029/2006WR00544.

18. De Lannoy, G.J.M., Reichle, R.H., Houser, P.R., Arsenault, K., Verhoest, N.E.C., Pauwels, V.R.N. (2011). Multi-Scale Assimilation of AMSR-E Snow Water Equivalent and MODIS Snow Cover Fraction in Northern Colorado, Water Resources Research, conditionally accepted.

19. Déry, S.J., V.V. Salomonson, M. Stieglitz, D.K. Hall and I. Appel, 2005. An approach to using snow areal depletion curves inferred from MODIS and its application to land surface modelling in Alaska. Hydrol. Processes, 19, 2755–2774.

20. Dirmeyer, P., 2000. Using a global soil wetness dataset to improve seasonal climate simulation. J. Climate, 13, 2900–2921.

21. Dong, J., and C. Peters-Lidard (2010). On the Relationship Between Temperature and MODIS Snow Cover Retrieval Errors in the Western U.S. IEEE Trans. Geosci. Remote Sens., 3(1), 132-140.

22. Dong, J., J.P. Walker and P.R. Houser, 2005. Factors Affecting Remotely Sensed Snow Water Equivalent Uncertainty. Remote Sensing of Environment, 97, 68-82, doi:10.1016/j.rse.2005.04.010.

23. Dong, J., J.P. Walker, P.R. Houser and C. Sun, 2007. Scanning multichannel microwave radiometer snow water equivalent assimilation. J. Geophys. Res., 112, D07108, doi:10.1029/2006JD007209.

24. Draper, C. S., J.-F. Mahfouf, and J. P. Walker (2011), Root zone soil moisture from the assimilation of screen-level variables and remotely sensed soil moisture, J. Geophys. Res., 116, D02127, doi:10.1029/2010JD013829.

25. Drusch, M. and Viterbo, P. (2007). Assimilation of screen-level variables in ECMWF's Integrated Forecast System: A study on the impact on the forecast quality and analyzed soil moisture, Monthly Weather Review, 135, 300-314.

26. Giustarini, L., Matgen, P., Hostache, R., Montanari, M., Plaza, D., Pauwels, V. R. N., De Lannoy, G. J. M., De Keyser, R., Pfister, L., Hoffmann, L, & Savenije, H. H. G. (2011). Assimilatign SAR-derived water level data into a flood models: a case study, Hydrol. Earth Syst. Sci., 15, 2349-2365.

27. Hostache, R., Matgen, P., Schumann, G., Puech, C., Hoffmann, L., & Pfister, L. (2009). Water level estimation and reduction of hydraulic model calibration uncertainties using satellite SAR images of floods, IEEE Transactions on Geoscience and Remote Sensing, 47, 431-441.

28. Houser, P., M.F. Hutchinson, P. Viterbo, J. Hervé Douville and S.W. Running, 2004. Terrestrial data assimilation, Chapter C.4 in Vegetation, Water, Humans and the Climate. Global Change - The IGB Series. Kabat, P. et al. (eds). Springer, Berlin, pp 273-287.

29. Houser, P.R., W.J. Shuttleworth, J.S. Famiglietti, H.V. Gupta, K.H. Syed and D.C. Goodrich, 1998. Integration of soil moisture remote sensing and hydrologic modeling using data assimilation. Water Resour. Res., 34, 3405-3420.

30. Hu, Y., X. Gao, W. Shuttleworth, H. Gupta and P. Viterbo, 1999. Soil moisture nudging experiments with a single column version of the ECMWF model. Q. J. R. Meteorol. Soc., 125, 1879–1902.

31. Kalman, R.E., 1960. A new approach to linear filtering and prediction problems. Trans. ASME, Ser. D, J. Basic Eng., 82, 35-45.

32. Koster, R.D., M. Suarez, P. Liu, U. Jambor, A. Berg, M. Kistler, R. Reichle, M. Rodell and J. Famiglietti, 2004. Realistic initialization of land surface states: Impacts on subseasonal forecast skill. J. Hydrometeorol., 5, 1049–1063.

33. Kostov, K.G. and T.J. Jackson, 1993. Estimating profile soil moisture from surface layer measurements – A review. In: Proc. The International

Society for Optical Engineering, Vol. 1941, Orlando, Florida, pp 125-136.

34. Liu, Q., Reichle, R.H., Bindlish, R., Cosh, M.H., Crow, W. T., de Jeu, R., De Lannoy, G.J.M., Huffman, G.J., Jackson, T.J. (2011). The contribution of precipitation forcing and satellite observations of soil moisture on the skill of soil moisture estimates in a land data assimilation system, Journal of Hydrometeorology, in press.

35. Lorenc, A., 1981. A global three-dimensional multivariate statistical interpolation scheme. Mon. Weather Rev., 109, 701-721.

36. Lorenc, A.C., R.S. Bell and B. Macpherson, 1991. The Meteorological Office analysis correction data assimilation scheme. Q. J. R. Meteorol. Soc., 117, 59-89.

37. Mahfouf, J.-F., Bergaoui, K., Draper, C., Bouyssel, C., Taillefer, F. and Taseva, L. (2009). A comparison of two off-line soil analysis schemes for assimilation of screen-level observations, Journal of Geophysical Research, 114, D08105.

38. Margulis, S.A., E.F. Wood and P.A. Troch, 2006. A terrestrial water cycle: Modeling and data assimilation across catchment scales. J. Hydrometeorol., 7, 309–311.

39. Matgen, P., Montanari, M., Hostache, R., Pfister, L., Hoffmann, L., Plaza, D., Pauwels, V. R. N., De Lannoy, G. J. M., De Keyser, R., & Savenije, H. H. G. (2010). Towards the sequential assimilation of SAR-derived water stages into hydraulic models using the Particle Filter: proof of concept, Hydrol. Earth Syst. Sci., 14, 1773-1785.

40. McLaughlin, D., 1995. Recent developments in hydrologic data assimilation. In U.S. National Report to the IUGG (1991-1994). Rev. Geophys., supplement, 977–984.

41. McLaughlin, D., 2002. An integrated approach to hydrologic data assimilation: interpolation, smoothing, and filtering. Adv. Water Resour., 25, 1275–1286.

42. Montanari, M., Hostache, R., Matgen, P., Schumann, G., Pfister, L., & Hoffmann, L. (2009). Calibration and sequential updating of a coupled hydrologic-hydraulic model using remote sensing-derived water stages, Hydrol. Earth Syst. Sci., 13, 367-380.

43. Nichols, N.K., 2001. State estimation using measured data in dynamic system models, Lecture notes for the Oxford/RAL Spring School in Quantitative Earth Observation.

44. Pan, M. and E.F. Wood, 2006. Data assimilation for estimating the terrestrial water budget using a constrained ensemble Kalman filter. J.

Hydrometeorol., 7, 534–547.

45. Parrish, D. and J. Derber, 1992. The National Meteorological Center's spectral statistical interpolation analysis system. Mon. Weather Rev., 120, 1747-1763.

46. Reichle, R.H. and R. Koster, 2003. Assessing the impact of horizontal error correlations in background fields on soil moisture estimation. J. Hydrometeorol., 4, 1229–1242.

47. Reichle, R.H. and R. Koster, 2004. Bias reduction in short records of satellite soil moisture. Geophys. Res. Lett., 31, L19501.1–L19501.4.

48. Reichle, R.H., D.B. McLaughlin and D. Entekhabi, 2002a. Hydrologic data assimilation with the ensemble Kalman filter. Mon. Weather Rev., 120, 103–114.

49. Reichle, R.H., J.P. Walker, P.R. Houser and R.D. Koster, 2002b. Extended versus ensemble Kalman filtering for land data assimilation. J. Hydrometeorol., 3, 728–740.

50. Reichle, R.H., W.T. Crow and C.L. Keppenne, 2008. An adaptive ensemble Kalman filter for soil moisture data assimilation, Water Resour. Res., 44, W03423, doi:10.1029/2007WR006357.

51. Rood, R.B., S.E. Cohn and L. Coy, 1994. Data assimilation for EOS: The value of assimilated data, Part 1. The Earth Observer, 6, 23-25.

52. Sahoo, A.K., De Lannoy, G.J.M., Reichle, R.H. (2011). Downscaling of AMSR-E soil moisture in the Little River Watershed, Georgia, USA. Advances in Water Resources, in review.

53. Sellers, P. J., Y. Mintz, and A. Dalcher, 1986: A simple biosphere model (SiB) for use within general circulation models. J. Atmos. Sci., 43: 505-531.

54. Slater, A.G. and M. Clark, 2006. Snow data assimilation via an ensemble Kalman filter. J. Hydrometeorol., 7, 478–493.

55. Stauffer, D.R. and N.L. Seaman, 1990. Use of four-dimensional data assimilation in a limited-area mesoscale model. Part I: Experiments with synoptic-scale data. Mon. Weather Rev., 118, 1250-1277.

56. Stieglitz, M., D. Rind, J. Famiglietti and C. Rosenzweig, 1997. An efficient approach to modeling the topographic control of surface hydrology for regional and global climate modeling. J. Climate, 10, 118–137.

57. Turner, M.R.J., J.P. Walker and P.R. Oke, 2007. Ensemble Member Generation for Sequential Data Assimilation. Remote Sensing of Environment, 112, doi:10.1016/j.rse.2007.02.042.

58. van Loon, E.E. and P.A. Troch, 2001. Directives for 4-D soil moisture data assimilation in hydrological modelling. IAHS, 270, 257-267.

59. Walker J.P. and P.R. Houser, 2001. A methodology for initialising soil moisture in a global climate model: assimilation of near-surface soil moisture observations, J. Geophys. Res., 106, 11,761-11,774.

60. Walker, J.P. and P.R. Houser, 2004. Requirements of a global near-surface soil moisture satellite mission: Accuracy, repeat time, and spatial resolution. Adv. Water Resour., 27, 785-801.

61. Walker, J.P. and P.R. Houser, 2005. Hydrologic data assimilation. In A. Aswathanarayana (Ed.), Advances in water science methodologies (230 pp). The Netherlands, A.A. Balkema.

62. Walker, J.P., G.R. Willgoose and J.D. Kalma, 2001a. One-dimensional soil moisture profile retrieval by assimilation of near-surface observations: A comparison of retrieval algorithms. Adv. Water Resour., 24, 631-650.

63. Walker, J.P., G.R. Willgoose and J.D. Kalma, 2001b. One-dimensional soil moisture profile retrieval by assimilation of near-surface measurements: A simplified soil moisture model and field application. J. Hydrometeorol., 2, 356-373.

64. Walker, J.P., G.R. Willgoose and J.D. Kalma, 2002. Three-dimensional soil moisture profile retrieval by assimilation of near-surface measurements: Simplified Kalman filter covariance forecasting and field application.

65. Water Resour. Res., 38, 1301, doi:10.1029/2002WR001545. Walker, J.P., P.R. Houser and R. Reichle, 2003. New technologies require advances in hydrologic data assimilation. EOS, 84, 545–551.

66. WMO, 1992. Simulated real-time intercomparison of hydrological models (Tech. Rep. No. 38). Geneva.

67. Zhang, H. and C.S. Frederiksen, 2003. Local and nonlocal impacts of soil moisture initialization on AGCM seasonal forecasts: A model sensitivity study. J. Climate, 16, 2117–2137.

Chapter 9

HYDROLOGICAL EFFECTS OF DIFFERENT SOIL MANAGEMENT PRACTICES IN MEDITERRANEAN AREAS

Giuseppe Bombino, Vincenzo Tamburino, Demetrio Antonio Zema and Santo Marcello Zimbone

Mediterranean University of Reggio Calabria Department of Agro-Forest and Environmental Sciences and Technologies Italy

INTRODUCTION

In Mediterranean environment intensive agricultural activities are often practiced in steep slopes, where sometimes climatic, geomorphologic and land use factors (e.g. the high rainfall intensity, the scarce vegetal coverage, especially on the occasion of the early rainfalls, the low organic matter content of soils, etc.) worsen the impacts of soil erosion. In such contexts agriculture may play an important role both in terms of economic and social spin-offs (e.g. peopling of hilly marginal lands) as well as under the environmental aspect (e.g. control of erosion phenomena). This is the case of olive growing practiced in hilly lands with a low tree density (e.g. in Southern Italy), often subjected to torrential rainstorms. Therefore, soil degradation problems in such agricultural steep lands under semi-arid conditions must be accounted for through proper soil management systems with low environmental impacts (mainly on soil hydrology).

Until recently, the most common practice for soil conservation in many Mediterranean regions, as Andalusia (Spain, Gomez et al., 2003) and Sicily or Calabria (Italy) has been tillage: however, the tradition of frequent tillage, aimed at preventing competition from natural vegetation for water and nutrients with the olive tree and at facilitating olive harvesting, has exacerbated the problems of erosion and soil degradation (Gomez et al., 2009a). Alternative practices to tillage include: no-tillage with herbicides to maintain a bare and weed-free soil (which sometimes results in accelerated soil erosion due to an increase in water runoff) or the use of a cover crop to protect the soil during autumn and winter, either sown in early autumn or from the regeneration of the natural vegetation after the onset of rains (Gomez et al., 2009a, 2009c).

The cover crop is controlled by mowing or by herbicide in spring to reduce the risk of competition for water with the trees, which represents the main limiting factor for plant growth in semi-arid lands, where the evapo-transpiration rate is very high and water resource is scarce.

Studies on soil erosion in orchards in Mediterranean environment have analyzed the hydrological effects of the traditional different managements systems (e.g. Dastgheib & Frampton, 2000; Gago et al., 2007; Gomez et al., 2003, 2009b, 2009c; Monteiro & Moreira, 2004); the important role for soil conservation played by the crop cover has been also highlighted, thanks to the rainfall interception and infiltrability increase (Kosmas et al., 1997; Gomez et al., 2003, 2009b, 2009c; Ramos & Martinez-Casasnovas, 2004).

In spite of the results achieved in these studies, the information about the impacts of different management practices on soil losses is still insufficient and does not allow a proper evaluation of the erosive risks in hilly olive groves across the different local conditions. This consideration is reflected in the contradictory results found in the literature. For example, Pastor et al. (1999) reported that, despite more rill erosion, no-tillage reduced soil losses as compared to conventional tillage, but Francia et al. (2000) measured the opposite effect in runoff plots. The few short term experiments mentioned cannot capture the long-term effects of soil management and, to the present knowledge, no previous work has attempted to assess systematically the effects of all soil management practices on soil losses in olive orchards (Gomez et al., 2003). These latter Authors argued as well that "the scarcity of experimental results is the bottleneck for improving the estimation of management effects on the rate of soil losses in olive plantations. Until additional field experiments measuring actual soil loss rates, and field surveys estimating historical rates of soil loss, are carried out at different conditions and scales, erosion rates will remain highly uncertain. On the other hand, qualitative observations indicate that the magnitude of the erosion problem in olive groves on steep slopes is such that the role of alternative soil management in limiting soil loss should be urgently assessed".

However, because of the high variability that characterizes the Mediterranean environments, soil erosion varies considerably over space and time and in most cases it is inappropriate to extrapolate these measures to other spatial units, where different hydrological and erosive processes take place (Taguas et al., 2010). Thus further detailed investigations also at plot scale could integrate literature data, in order to estimate in different contexts the magnitude of the erosive risk: this latter, considering that monitoring activities of surface runoff and soil loss are time consuming and expensive tasks, can be assessed also through a modeling approach by mathematical simulation of water runoff and soil erosion processes.

As well known, prediction models are useful tools for monitoring and controlling the impacts of soil erosion (e.g. Engel et al., 1993; Licciardello et al., 2007; Zema et al., 2011). While the potential of process based models is greater in comparison to empirical ones, their complexity means larger data requirements, potentially greater problems of error propagation and increased difficulty in understanding the way the model simulates the erosion processes (Favis-Mortlock et al., 2001). Published comparisons between the two types show that the average error and model efficiency in predicting soil loss are similar (Morgan & Nearing, 2000; Tiwari et al., 2000). Thus, empirical models, mainly the Universal Soil Loss Equation (USLE; Wischmeier & Smith, 1978) or its derivatives (e.g. RUSLE; Renard et al. 1997), are still widely used (Gomez et al., 2003): in fact, the reduced data requirement and simplicity of USLE-type models (compared to process-based ones) make them useful tools for planning activities destined to soil conservation workers (e.g. Taguas et al., 2010).

Such considerations have stimulated research activities to evaluate and predict the erosion risks in hilly olive groves of Calabria region (Southern Italy), where olive growing represents a fundamental sector of local economy and the most important land use. Within such research activities, this paper aims at: (i) integrating the literature data on the hydrologic effects of three soil management practices (conventional tillage, no tillage and crop cover) typical of the Mediterranean olive groves; (ii) drawing indications on erosion prediction capability of the RUSLE model for the experimental conditions.

MATERIALS AND METHODS

The Study Area

The study area is located on the northern side of the torrent Menga valley near Gallina di Reggio Calabria in Southern Italy (Figure 1). The site lies at an altitude of approximately 250 m above sea level; predominant aspect is south. Soil has been classified as sandy-loam (USDA SCS, 1984). The climate of the area is typically Mediterranean, with a mean yearly precipitation of ca. 600 mm, most of which are concentrated in fall and winter periods. Mean monthly temperatures range from 11.5 °C in January, which is the coldest month in the year, to 26.5 °C in July (Bombino et al., 2004).

The Experimental Design

In 1991 a research group of the University of Reggio Calabria established nine experimental plots at the site (Figure 2), in order to monitor runoff and soil erosion under different slope and vegetation conditions (Bombino et al., 2002). The plots were characterized by different lengths and slope; the three longer (33 m) plots had a 9% slope, whereas three of the six shorter (22 m) plots had a 9% slope and three an 18% slope. A sheet metal cutoff wall, fixing 30 cm into the soil and protruding 20 cm above the ground surface, was installed around the upper and the two adjacent sides of each plot in order to hydrologically isolate the plots. On the lower side of each plot, a 1-m³ tank was installed to collect runoff volumes and sediment loads. Rainfall has been recorded at the site since 1990, using a tipping bucket rain gauge, but measurements of runoff and sediment concentrations from the plots have been available since February 2002. Monthly and after each storm event, the sediment load collected in the tank was well mixed and several 1-liter suspended sediment samples were taken from different depths within the tank. The sediment concentration in each sample was determined by oven drying at 105 °C and the mean value of the samples was calculated. The sediment load from each plot was then calculated as the product of the mean sediment concentration and the water volume measured in the tank.

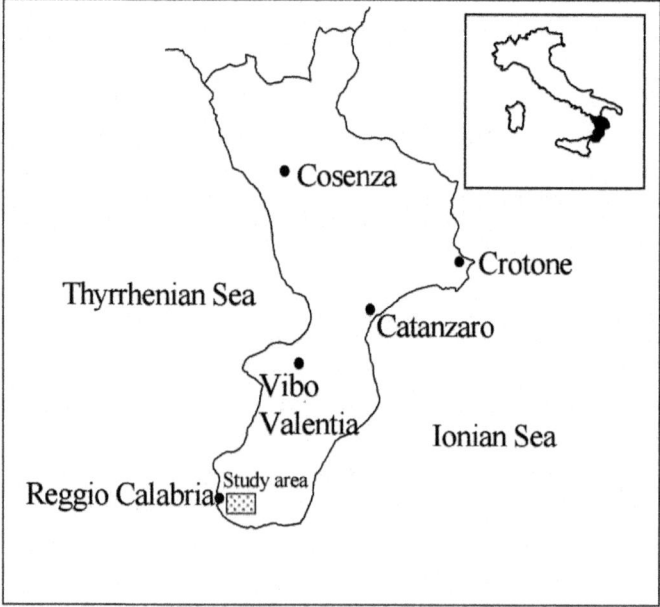

Figure 1. Location of the study area.

In order to determine the plot vegetal coverage, monthly surveys have been performed in each plot since October 2001. The canopy cover of herbaceous and shrub layers (in %) was evaluated within 1 x 1-m^2 sample areas (at least 1 every 25 m^2).

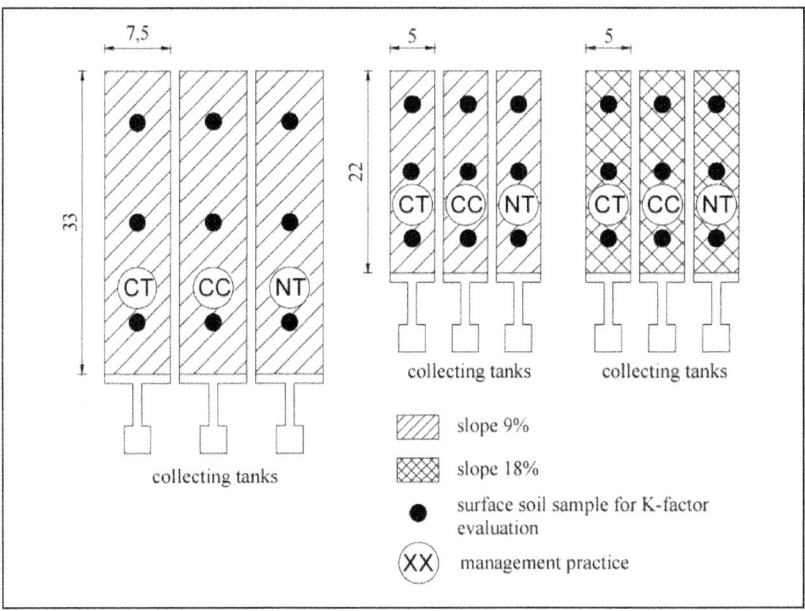

Figure 2. Layout of the experimental plots (linear measures in metres) (CT = Conventional Tillage; NT = No Tillage; CC = Crop Cover).

Evaluation of the Soil Management Practices

In the experimental plots three soil management practices commonly adopted in hilly olive groves of the Mediterranean areas were simulated and their hydrological effects were measured and compared. Conventional tillage (hereafter CT) and total weed killing through herbicide (hereafter no tillage, NT) were compared with a conservative practice (hereafter crop cover, CC) based on Low Dosage Herbicide Treatments (LDHT).

CT consisted of two to three passes, 10-15 cm deep, with a trough milling machines with subsequent soil compaction, generally starting after the first rain in October to control weeds in the whole plot. NT consisted of maintaining the soil weed-free and bare with 3 to 4 herbicide applications (acid glyphosate at a dose of 2.1 kg a.e. ha^{-1} distributed manually or through a backpack sprayer) per year, mostly concentrated in late autumn (November-December). In CC practice the plots were subjected to two herbicide treatments in October and

April (acid glyphosate at a dose of 0.23 kg a.e. ha⁻¹ distributed manually or through a backpack sprayer). For all thesis runoff volumes and sediment concentrations were measured during a 7-year monitoring period (February 2002-December 2008); such values together with calculated soil losses were aggregated at monthly and yearly scales and then averaged among the plots subjected to the same experimental soil management practice (Figure 2).

Implementation of the RUSLE Model

The RUSLE model was implemented at yearly scale in order to verify its prediction capability of soil erosion for the investigated soil management practices.

To calculate the R-factor, a simple equation correlating the erosivity index for the e-th event (R_e, MJ mm ha⁻¹ h⁻¹) and the corresponding rainfall height (he, mm) was utilized, due to the unavailability of rainfall records at sub-hourly scale in the meteorological database:

$$R_e = \alpha h_e^{\beta},$$
(1)

a and b are empirical coefficients for which the values of 0.18 and 1.59 respectively, calculated for the very close meteorological station of Messina (Bagarello & D'Asaro, 1994), were assumed (Table 1).

The K-factor (0.65 t ha⁻¹ per R-factor unit, Table 1) was averaged from K factors established for several soil samples collected within the investigated plots (Figure 2).

Topographic factor values $L_i S_i$ for the i-th plot were calculated by using the following relationship (McCool et al., 1989) (Table 1):

$$L_i S_i = \left(\frac{\lambda_i}{22.13}\right)^{m_i} \left(16.8 \; \sin\alpha_i - 0.5\right),$$
(2)

where λ_i (m) is the slope length of the i-th plot, α_i is the slope angle (Figure 2); m_i was calculated as follows:

$$m_i = \frac{f_i}{\left(1 + f_i\right)}$$
(3)

being:

$$f_i = \frac{\sin\alpha_i}{0.0896 \left(3 \; \sin 0.8 \; \alpha_i + 0.56\right)}.$$
(4)

Because the C-factor changes continuously with cover and residue among cutting operations, the related values need to be established for different periods during the year, according the guidelines of Wischmeier & Smith (1978). Therefore, the monthly C-factors for the three soil management practices (Table 1) were calculated as a function of the plot vegetal coverage (reported for the investigated soil management practices in Table 2) through a regression equation ($r^2 = 0.98$; $n = 6$), correlating the C-values - in the range 0.0032-0.45, reported by Bazzoffi (2007) for vegetated or unvegetated olive orchards - to the corresponding per cent vegetal coverage.

Table 1. Value or range of RUSLE factors for the experimental plots.

RUSLE factor	Value or range
Max R_e (MJ mm ha^{-1} h^{-1})	416.96
K (t ha^{-1} h t^{-1} m^{-1} mm^{-1} ha)	0.65
LS (-)	1.0 to 2.45
C (-)	0.01 to 0.40
P (-)	0.6 to 1.0

According to the guidelines of Wischmeier & Smith (1978), the P-factor was assumed equal to 0.6 (slope of 9%) or 0.8 (slope of 18%) in occurrence of tillage operations along contour lines for CT; otherwise a value of 1.0 was considered, because no erosion control practice was adopted (Table 1).

Table 2. Monthly values of plot vegetal coverage for the investigated soil management practices (CT = Conventional Tillage; NT = No Tillage; CC = Crop Cover).

Month	Plot vegetal coverage (%)		
	CT	NT	CC
January	25.9	16.7	58,1
February	36.8	21.8	71,3
March	40.9	29.5	82,2
April	52.8	33.0	22,4[1]
May	57.1	34.2	43,4
June	54.5	32.8	45,6
July	48.3	25.6	41,1
August	32.1	19.8	36,9
September	36.5	21.2	44,3
October	49.6	26.7	19,4[1]
November	2.3[1]	9.8[1]	33,9
December	10.8	11.2	47,1

[1] Treatment date

Evaluation of the RUSLE Model

Model performance was evaluated at yearly scale by qualitative and quantitative approaches. The qualitative procedure consisted of visually comparing observed and simulated values. For quantitative evaluation a range of both summary and difference measures were used.

The summary measures utilized were the mean and standard deviation of both observed and simulated values. Given that coefficient of determination (r^2) is an insufficient and often misleading evaluation criterion (Licciardello et al., 2007; Zema et al., 2011), the Nash & Sutcliffe (1970) coefficient of efficiency (E) was also used to assess model efficiency. As suggested by Krause et al. (2005) and Legates & McCabe (1999), E was integrated with the Root Mean Square Error (RMSE), which describes the difference between the observed values and the model predictions in the unit of the variable. The values considered to be optimal for these criteria were 1 for r^2 and E and 0 for RMSE. According to common practice, simulation results are considered good for values of E greater than or equal to 0.75, satisfactory for values of E between 0.75 and 0.36, and unsatisfactory for values below 0.36 (Van Liew & Garbrecht, 2003).

RESULTS AND DISCUSSION

Evaluation of the Soil Management Practices

The results of the comparison among the hydrologic effects of the three investigated soil management practices (conventional tillage, no tillage and crop cover), typical of Mediterranean hilly olive groves, are reported in this section.

Analysis at Yearly Scale

There was a clear difference in the runoff volumes yielded in the investigated management practices, with NT having the highest runoff coefficient and CC the lowest. The LDHT (CC practice) produced average surface runoff volumes lower by 28% than CT; conversely, complete removal of vegetal coverage through herbicide (NT) resulted in average runoff volume higher by 28% and 79% than CT or CC respectively. Consequently mean yearly values of the runoff coefficient for CC (10.7%) were appreciably lower than those recorded for CT (15.0%) and NT (19.4%) practices (Table 3).

The differences among the soil management practices in the average yearly runoff coefficients measured in this study are basically coherent with those measured in other investigations available in literature. Raglione et al. (1999)

reported runoff coefficients of 3.5 and 12.8% for CC and CT respectively in Calabria (southern Italy). Bruggeman et al. (2005) measured average runoff of 184 and 66.5 mm year[1] for orchards under CT and CC, respectively, in Syria, in an area with an average yearly precipitation of 400 mm year[1]. Francia et al. (2006) measured, in a loamy soil on a 30% slope, higher runoff coefficients in the treatment under NT (5.3%) and lower values for CT and CC (1.5 and 2.7%) respectively. Gomez & Giraldez (2007), in a sandy-loam soil on a 11% slope, measured runoff coefficients of 20 and 5.7% for CT and CC respectively. More recently, in Andalusia (Spain) Gomez et al. (2009b) in a 4-year experiment carried out in an olive tree farm on a sandy-loam soil found runoff coefficients of 6 and 16% for CC and CT practices respectively; in the same environment, Gomez et al. (2009c) recorded during a 7-year experiment in a young olive grove installed on a heavy clay soil the highest average yearly runoff coefficient (11.9%) for NT, which decreased to 1.2% for CC and to 3.1% for CT.

Sediment concentration in collected runoff samples was lower for plots subjected to LDHT (54% less than in CT plots) and higher for NT treatment (18% higher than CT) (Table 3).

The advantages induced by application of low doses of herbicide (CC) were particularly remarkably in terms of soil loss, decreased in this soil management practice by 57% and 71% with respect to CT and NT (Table 3). As well known, the soil loss depends not only on the runoff generation, but also on the sediment concentration of the water stream; both were greater under CT and NT treatments, which left for some periods along the year the soil unprotected and then exposed to the erosion risk. The records of the yearly soil losses for the three experimental soil management practices show a large inter-annual variability, with average values of 28.8 t ha[-1] year[-1] (with a standard deviation of 34.1 t ha[-1] year[-1]) in the CT practice and 42.2 t ha[-1] year[-1] (± 50.0 t ha[-1] year[-1]) under NT with the lowest average value recorded for CC (12.3 ± 14.7 t ha[-1] year[-1]) (Table 3).

In all the observation years CC practice allowed to achieve soil losses very close to the tolerable value of 11-12 t ha[-1] year[-1] suggested by several Authors (e.g. Montgomery, 2007; Stone et al., 2000); conversely under CT and NT treatments such a threshold was always exceeded (Table 3).

A comparison between the yearly soil losses measured during the 7-year monitoring period of the present study and the values reported by other Authors in experimental runoff plots to evaluate soil erosion in olive groves has been carried out; the main results are reported in Table 4

Kosmas et al. (1996) measured soil losses between 0 and 0.03 t ha[-1] year[-1] in semi-natural olive groves in Greece with 90% of the soil covered by vegetation. Raglione et al. (1999) measured in Calabria total soil losses of 0.36

and 41 t ha^{-1} year^{-1} for CC and CT respectively in a 2-year plot experiment. In Syria Bruggeman et al. (2005) measured average soil losses of 11.2 and 41.4 t ha^{-1} year^{-1} in orchards under CC and CT respectively in an area with a slope of 24% for a 4-year period.

Table 3. a, b. Yearly values of the hydrological observations for the investigated soil management practices (CT = Conventional Tillage; NT = No Tillage; CC = Crop Cover) in the experimental plots.

Year	Rainfall (mm)	Cumulated surface runoff (mm)			Runoff coefficient (%)		
		CT	NT	CC	CT	NT	CC
2002[1]	689.2	105.8	155.2	86.0	15.4	22.5	12.5
2003	843.3	136.5	180.2	103.2	16.2	21.4	12.2
2004	522.2	72.5	105.6	52.4	13.9	20.2	10.0
2005	690.4	113.4	120.6	76.4	16.4	17.5	11.1
2006	521.4	71.0	89.0	47.5	13.6	17.1	9.1
2007	690.4	113.4	120.6	76.4	16.4	17.5	11.1
2008	622.0	82.2	120.6	56.2	13.2	19.4	9.0
Cumulated	4578.9	694.7	891.8	498.1		-	
Mean[2]	654.1	99.2[a]	127.4[a]	71.2[b]	15.0[a]	19.4[b]	10.7[c]

(a)

Year	Rainfall (mm)	Mean sediment concentration (g l^{-1})			Cumulated soil loss (t ha^{-1})		
		CT	NT	CC	CT	NT	CC
2002[1]	689.2	14.1	12.4	3.7	28.4	47.2	8.8
2003	843.3	6.0	5.6	3.6	40.9	52.9	17.3
2004	522.2	6.4	6.5	5.7	19.0	28.4	10.1
2005	690.4	6.7	8.1	4.4	32.2	44.6	14.8
2006	521.4	14.9	12.6	5.7	18.2	29.5	6.8
2007	690.4	6.3	8.4	4.7	32.2	44.6	14.8
2008	622.0	7.9	19.7	1.1	30.2	48.3	13.3
Cumulated	4578.9		-		201.3	295.4	86.0
Mean[2]	654.1	8.9[a]	10.5[a]	4.1[b]	28.8[a]	42.2[b]	12.3[c]

(b)

[1]February-December

[2]Values followed by the same letter are not significantly different at $P < 0.05$.

Gomez et al. (2004) reported average soil losses of 4.0, 8.5 and 1.2 t ha^{-1} year^{-1} from CT, NT and CC in a 3-year experiment on a heavy clay soil in Andalusia. In a 2-year study carried out n the same region, Francia et al. (2006) measured soil losses of 5.7, 25.6 and 2.1 t ha^{-1} from CT, NT and CC respectively. Also in Andalusia, Gomez & Giraldez (2007) reported average soil losses of 21.5 and 0.4 t ha^{-1} year^{-1} for CT and CC in a different 4-year experiment. More recently, Gomez et al. (2009c) in a 7-year study reported soil losses of 2.9 t ha^{-1} year^{-1} for CT, 6.9 t ha^{-1} year^{-1} for NT and 0.8 t ha^{-1} year^{-1} for CC in a young olive grove installed on a heavy clay soil of Andalusia; in the same environment, Gomez et al. (2009b) in a 4-year experiment carried out in an olive tree farm on a sandy-loam soil recorded soil losses of 1.9 and 0.4 t ha^{-1} year^{-1} for CT and CC treatments respectively. Average soil losses measured in our experimental plots subjected to CT management practice (28.8 t ha^{-1} year^{-1}) are coherent with the studies by Raglione et al. (1999), Bruggeman et al. (2005) and Gomez & Giraldez (2007), but generally higher than the observations reported in the other investigations (Francia et al., 2006; Gomez et al., 2004; Gomez et al., 2009b, 2009c). Also soil losses observed in the present study for NT and CC soil management practices (42.2 and 12.3 t ha^{-1} year^{-1} respectively) were generally higher than the observations found in the mentioned studies, except for data reported by Bruggeman et al. (2005) for CC, which are very close to the value achieved in the present study (Table 3). Even though the comparison of these values must be made with care due to relevant variability in the experimental climatic, morphological and management conditions among the examined studies and the limited duration of many of these databases (at most 4 years), the magnitude of the soil losses achieved in the present study highlighted the severity of the erosion phenomena in the experimental conditions and, as a consequence, the need of countermeasures to control and mitigate the erosive risks.

Table 4. Soil losses in experimental plots to evaluate soil erosion in olive groves reported in the available literature.

Study area	Authors	Soil losses (t ha^{-1} year^{-1})		
		CT	NT	CC
Calabria, Italy	Present study	28.8	42.2	12.3
	Raglione et al. (1999)	41.0	-	0.36
Andalusia, Spain	Gomez et al. (2004)	4.0	8.5	1.2
	Francia et al. (2006)	5.7	25.6	2.1
	Gomez & Giraldez (2007)	21.5	-	0.4
	Gomez et al. (2009b)	1.9	-	0.4
	Gomez et al. (2009c)	2.9	6.9	0.8
Syria	Bruggeman et al. (2005)	41.4	-	11.2
Greece	Kosmas et al. (1996)	0 to 0.03		

Analysis at monthly scale Figures 3 a, b and c illustrate the values (aggregated or averaged for 3-month periods) of surface runoff, sediment concentration and soil loss achieved in the experimental plots during the monitoring period. It is evident the remarkable reduction of all the hydrological variables recorded in the plots subjected to LDHT in comparison with the other soil management practices (and particularly with NT treatment). Gomez et al. (2009c) remarked a general reduction of runoff for all the hydrological variables along the monitoring period as the experiment progressed, contrary to what found in our experimental plots.

The analysis made at monthly scale highlighted that runoff was mainly concentrated from October to March, i.e. in the months characterized by the highest rainfalls and when the soil was moist after the dry season.

Table 5. a, b. Mean monthly values of the hydrological observations for the investigated soil management practices (CT = Conventional Tillage; NT = No Tillage; CC = Crop Cover) in the experimental plots.

Month	Rainfall (mm)	Surface runoff (mm)			Runoff coefficient (%)		
		CT	NT	CC	CT	CT	CT
January[1]	47.3	8.2	10.8	6.7	17.3	22.9	14.3
February	47.5	7.4	9.8	6.0	15.7	20.7	12.7
March	60.7	8.8	11.7	6.9	14.5	19.2	11.4
April	46.7	6.0	7.7	5.0	12.9	16.5	10.8
May	31.5	3.7	4.5	2.7	11.8	14.4	8.7
June	22.6	1.3	1.7	0.9	5.8	7.6	3.9
July	20.7	1.3	2.6	1.2	6.4	12.4	5.7
August	15.3	0.2	0.6	0.1	1.1	4.0	0.7
September	52.2	5.3	8.4	3.6	10.1	16.2	6.9
October	77.7	13.9	16.6	9.1	17.9	21.3	11.7
November	69.0	12.6	15.3	9.5	18.3	22.2	13.7
December	128.1	26.4	35.2	18.3	20.6	27.5	14.2

(a)

Month	Rainfall (mm)	Sediment concentration (g l⁻¹)			Soil loss (t ha⁻¹)		
		CT	NT	CC	CT	NT	CC
January[1]	47.3	9.3	28.5	2.3	2.1	3.3	1.0
February	47.5	11.1	14.5	8.3	2.7	4.5	1.2
March	60.7	8.2	7.9	2.5	2.7	3.4	1.0
April	46.7	7.8	8.8	3.7	2.1	2.8	0.9
May	31.5	7.8	9.2	2.2	1.6	2.2	0.4
June	22.6	2.1	2.3	1.5	0.7	0.9	0.3
July	20.7	9.61	5.0	1.3	1.5	1.9	0.2
August	15.3	16.4	11.5	4.3	0.2	0.3	0.1
September	52.2	7.2	6.2	4.2	1.7	2.5	0.7
October	77.7	5.6	5.9	2.6	3.3	4.6	1.5
November	69.0	5.5	6.7	3.3	2.4	3.8	1.1
December	128.1	5.3	6.6	3.1	5.5	8.3	2.5

(b)

[1]The mean values of January are calculated for the years 2003-2008

Soil losses recorded under CC were systematically lower than under other soil management practices, particularly in the late autumn-winter-early spring (up to 60% and 72% less than CT and NT treatments respectively), when rainfall erosivity was higher; this is attributable to the reduction of both surface runoff and sediment concentration, linked to the higher vegetal coverage (in the range 33.9-82.2% of the plot area, Table 2), which helped to reduce soil erosion. In fact, the herbicide application at low doses assured the survival of some spontaneous species (represented mainly by Crepis versicaria, Reichardia picroides, Inula viscosa, Salvia verbenacea, Oxalis pescapre, Arundo donax, Cynodon dactylon, Hedysarum coronarium, Foeniculum vulgare and Verbascum simatum) and the presence of biomass residues (consisting of the depressed species laid on the soil) during the wettest months, shielding wide portions of soil from the erosive impact of rainfall.

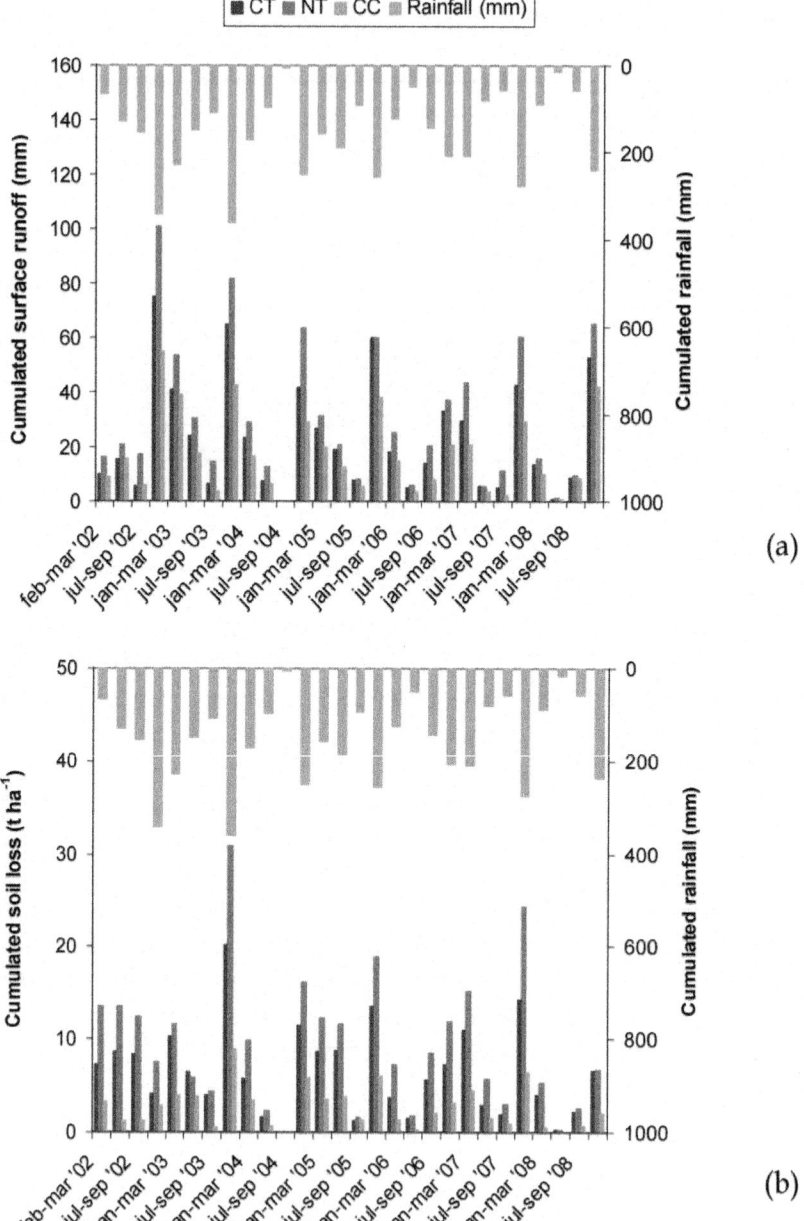

(a)

(b)

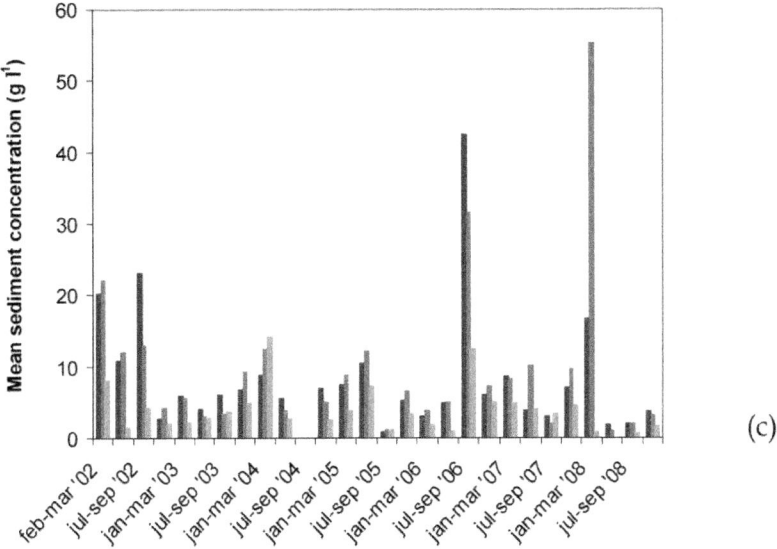

(c)

Figure 3. a, b, c. 3-month values of the hydrological observations for the investigated soil management practices (CT = Conventional Tillage; NT = No Tillage; CC = Crop Cover) in the experimental plots.

Conversely, total weed killing through herbicide (NT treatment), which destroyed crop residues, exposed the bare soil to the rainfall erosivity and thus to the erosion risks. In the summer months, characterized by low values of rainfall erosivity, the decay effects of weeds due to LDHT remarked since April helped to reduce competition for water between weeds and crop trees.

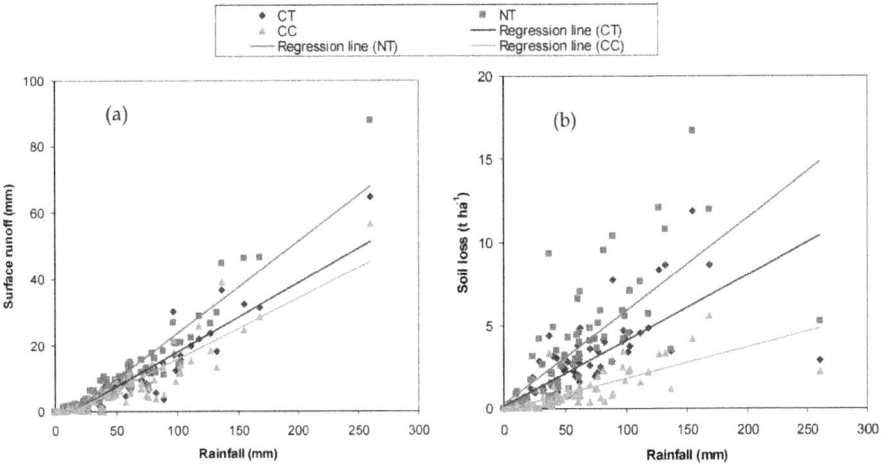

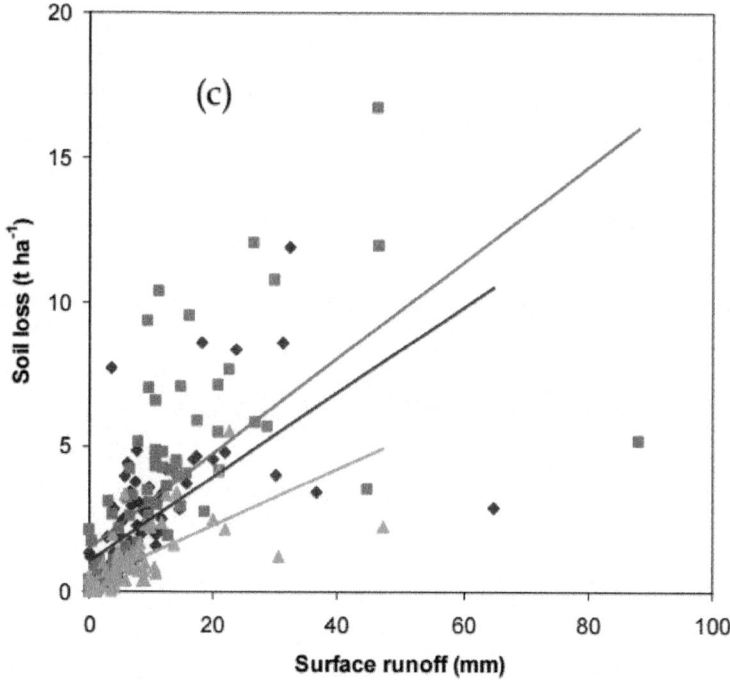

Figure 4. a, b, c. Linear regressions among monthly hydrological observations for the investigated soil management practices (CT = Conventional Tillage; NT = No Tillage; CC = Crop Cover) in the experimental plots.

The highest reduction of soil erosion was recorded for CC in December (which is characterized by the highest mean rainfall amount), when a soil loss lower by over 55% than in the other soil management practices was achieved (Table 5).

As expected, monthly runoff volumes were well correlated with the corresponding rainfalls (r^2 always higher than 0.83, with the maximum value of 0.89 achieved for NT treatment).

Lower values ($r^2 = 0.58$-0.62) were found in the regression relationships between monthly rainfall and soil loss. Finally the latter was weakly correlated with the corresponding runoff ($r^2 = 0.41$-0.47, Table 6), highlighting that sediment losses generally did not follow the same patterns as runoff volumes (Figure 4).

On the whole, LDHT led to average soil losses lower by about 60-70% than in the other soil management practices investigated in the present study. Reduced soil losses depended not only on the lower runoff volumes (presumably due to the increased interception induced by the wider vegetal cover, to the higher soil infiltration capacity and to the greater flow resistance linked to the presence of vegetation stems, which helps to dissipate water stream energy), but also on the lower sediment concentration (Tables 2 and 5). These positive effects seem to influence erosion rates more efficiently than CT treatment, which in its turn increases the water retention within surface hollows left by tillage (due to the increased soil roughness) or infiltration capacity induced by the higher soil surface porosity in comparison with NT treatment.

Table 6. Coefficients of linear regression among monthly hydrological observations for the investigated soil management practices (CT = Conventional Tillage; NT = No Tillage; CC = Crop Cover) in the experimental plots.

Soil management practice	Runoff-rainfall	Soil loss-rainfall	Soil loss-runoff
CT	0.87	0.60	0.41
NT	0.89	0.58	0.43
CC	0.83	0.62	0.47

The results of the present study are consistent with the other similar experiences aiming at evaluating the effects of some management practices on soil erosion: such studies in general suggest to adopt CC practice in olive groves, which, establishing a proper vegetal coverage of the soil in olive grove lanes, thus reduces runoff volumes and, as a consequence, soil losses more efficiently than the most common CT treatment. No tillage should be avoided, due to the fact that keeping the soil bare by herbicide application just in the months characterized by the highest rainfall erosivity (i.e. in late autumn, early spring or during the winter) reduces soil infiltration capacity and roughness, increasing water runoff and stream velocity as well as yielding the maximum erosion rates.

Evaluation of the RUSLE Model

The comparison among the soil losses measured in the experimental plots and the corresponding values predicted by the RUSLE model highlighted an unsatisfactory prediction capability at yearly scale. It is shown by the low coefficients of determination and efficiency as well as the high RMSE; also the differences between the measured and predicted standard deviations were high (Tables 7 and 8; Figure 5). The RUSLE model tended to overestimate soil losses for CT and, particularly, CC; on the contrary, soil losses measured for NT soil management practice were slightly underestimated (Figure 5).

For two (CT and NT) of the three simulated soil management practices the mean values of the predicted soil losses were close to the corresponding measured values with differences lower than 7%; also the differences between the measured and predicted cumulated soil losses, calculated for the entire 7-year monitoring period (201.3 versus 211.9 t ha^{-1} year^{-1} for CT treatment and 295.4 versus 273.8 t ha^{-1} year^{-1} for NT), were low. For CC soil management practice mean and total soil losses measured in the experimental plots and predicted by the RUSLE model differed instead by about 75-80% (Table 7). It means that, at least for the experimental conditions, estimations of soil losses performed by the RUSLE model must be considered with care, due to the fact that RUSLE is mainly meant to be used for long-term estimates of soil loss (Shrestha et al., 2006; Yoder et al., 2001).

Table 7. Yearly and cumulated values of soil losses measured in the experimental plots and predicted by the RUSLE model for the investigated soil management practices (CT = Conventional Tillage; NT = No Tillage; CC = Crop Cover).

| Year | Soil loss (t ha^{-1} year^{-1}) | | | | | |
| | Measured | | | Predicted | | |
	CT	NT	CC	CT	NT	CC
2002	28.4	47.2	8.8	42.4	66.2	21.4
2003	40.9	52.9	17.3	56.1	61.0	37.3
2004	19.0	28.4	10.1	30.1	37.6	11.0
2005	32.2	44.6	14.8	24.7	29.9	25.2
2006	18.2	29.5	6.8	18.8	25.6	15.3
2007	32.2	44.6	14.8	18.2	27.5	20.8
2008	30.2	48.3	13.3	21.6	26.0	19.1
Cumulated	201.3	295.4	85.9	211.9	273.8	150.1

Table 8. Statistics, efficiency and difference indexes of the RUSLE model at yearly scale for the investigated soil management practices (CT = Conventional Tillage; NT = No Tillage; CC = Crop Cover) in the experimental plots.

Soil management practice	Soil loss	Mean (t ha⁻¹ year⁻¹)	Std. Dev. (t ha⁻¹ year⁻¹)	r^2	E	RMSE (t ha⁻¹ year⁻¹)
CT	Measured	28.8	8.0	0.26	-1.11	10.70
	Predicted	30.6	13.1			
NT	Measured	42.2	9.5	0.14	-1.27	13.25
	Predicted	39.9	14.5			
CC	Measured	12.3	3.8	0.57	-10.04	11.65
	Predicted	22.0	9.3			

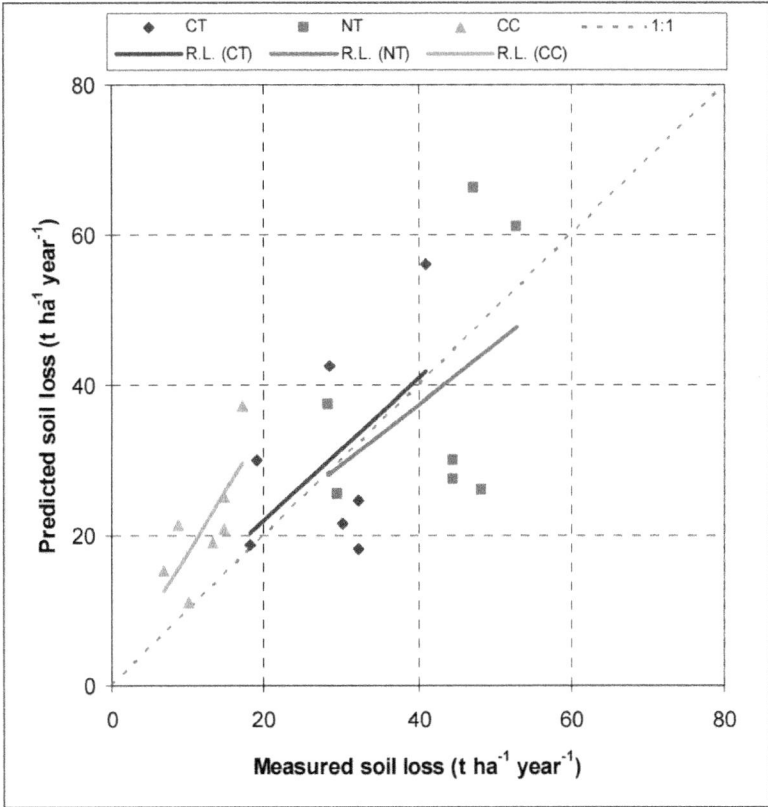

Figure 5. Comparison between soil loss measured in the experimental plots and predicted at yearly scale by the RUSLE model for the investigated soil management practices (CT = Conventional Tillage; NT = No Tillage; CC = Crop Cover) (R.L. indicates the regression lines).

It cannot be excluded that the unsatisfactory prediction capability of soil loss shown at yearly scale by the RUSLE model for the experimental plots can be attributable to:

- the unavailability of rainfall records at sub-hourly scale in the meteorological database, which, as mentioned above, forced the modeler to turn to an empiric regression equation for calculating RUSLE R-factors;

- the uncertainty of the calculated values of the C-factor (which is perhaps the most important USLE factor, because it represents conditions that can be managed most easily to reduce erosion, Ferreira et al., 1995; Renard et al., 1991; Renard & Ferreira, 1993; Yoder et al., 2001); it comes from the fact that the available soil management database lacked some important parameters (e.g. surface roughness and soil moisture) which can strongly influence soil loss estimation performed through the RUSLE model.

CONCLUSIONS

The present investigation has evaluated and simulated at plot scale the hydrological effects of three different soil management practices (conventional tillage, no tillage and crop cover through LDHT), commonly adopted in hilly olive groves of the Mediterranean environment. Although the monitoring of the erosion risk carried out in this paper is based on only 7 years of data and therefore results may change over a longer time period, the findings of this investigation highlight that, under the experimental conditions, the soil losses recorded for CT and NT practices are of high magnitude and thus unsustainable to avoid land degradation. Conversely, although the erosion rates achieved for CC practice in this study are generally higher than the observations reported by other Authors, LDHT, allowing to keep soil losses close to the tolerable value of 11-12 t ha^{-1} year^{-1} suggested by some Authors (e.g. Montgomery, 2007; Stone et al., 2000), results in a more efficient conservation practice in comparison to CT and NT and represents a valid alternative to these soil conservation practices. As a matter of fact, LDHT, assuring a suitable soil coverage during wet periods and a greater water availability to the olive trees in the dry seasons (thus reducing water competition with weeds), allows to mitigate the erosion risks and avoids negative impacts on crop productivity.

Unfortunately, also farmers operating in southern Italy, as remarked by other Authors (Gomez et al., 2009b; Helling & Haigh, 2002) in their respective countries, are in general reluctant to adopt soil management practices assuring a suitable crop cover and then high hydrological benefits during the wettest months (as LDHT), especially if they do not represent an immediate increase in the crop yield. Gomez (2005, 2009b) argued that the reasons for this reluctance is the need for a careful management of the cover crop to avoid competition for water with the olive tree (which is however basically limited) as well as the lower cost, for many farmers, of tillage (especially surface tillage) in comparison to cover crop soil management. This suggests the need of information activities by experts of soil conservation and farm advisers, purposing at illustrating the environmental benefits of cover crop soil management in olive groves, in particular: (i) immediately after olive planting; (ii) in young olive groves; or even (iii) in mature plantations with a very low tree density (especially in steep lands), where the canopy cover is low and the interception is rather limited. Thus, this kind of investigations may help to improve the countermeasures against soil erosion in Mediterranean slope zones, encouraging farmers to adopt soil conservation practices also through proper criteria of public financial support.

On a modeling approach, the present study has highlighted that the utilization of the RUSLE erosion model under the experimental conditions must be done with care, given that soil loss estimations have been reliable only for CT and NT treatments at a multi-year scale; presumably, a more complete hydrologic and geomorphologic database could improve model predictions.

Even though the outcomes of this study might contribute to soil conservation through sustainable management systems in agricultural lands characterized by high erosion risks, further research activities are finally needed not only to validate these results under different geomorphologic conditions, but also to assure a better understanding of runoff and erosion processes and to predict its effects with time.

REFERENCES

1. Bagarello, V. & D'Asaro, F. (1994). Estimating single storm erosion index. Transactions of the ASABE, Vol. 37, No. 3, pp. 785-791.
2. Bazzoffi, P. (2007). Universal Soil Loss Equation Calculator (USLE CALCULATOR). In: Erosione del suolo e sviluppo rurale, P. Bazzoffi (Ed.), Edagricole, Bologna, Italy, pp. 264 (in Italian).
3. Bombino, G., Porto, P. & Zimbone, S.M. (2002). Evaluating the crop and management factor C for applying RUSLE at plot scale. Proceedings of the 95th Annual International Meeting of ASAE held jointly with XVth World Congress of CIGR, Chicago, USA, 28-31 July 2002.
4. Bombino, G., Porto, P., Tamburino, V. & Zimbone, S.M. (2004). Crop and management factor estimate for applying RUSLE in rangeland

areas. Proceedings of the ASAE Annual International Meeting, Ottawa, Ontario, Canada, 1-4 August 2004.

5. Bruggeman, A., Masri, Z., Turkelboom, F., Zobisch, M. & El-Naheb, H. (2005). Strategies to sustain productivity of olive groves on steep slopes in the northwest of the Syrian Arab Republic. In: Integrated Soil and Water Management for Orchard Development; Role and Importance, Benites, J., Pisante, M. & Stagnari, F. (Eds.), FAO Land and Water Bulletin, vol. 10, FAO, Rome, Italy, pp. 75–87.

6. Dastgheib, F. & Frampton, C. (2000). Weed management practices in apple orchards and vineyards in the South Island of New Zealand. New Zealand Journal of Crop and Horticultural Science, Vol. 28, pp. 53–58.

7. Engel, B.A., Srinivasan, R. & Rewerts, C. (1993). A spatial decision system for modelling and managing agricultural non-point-source pollution. In: Environmental modelling with GIS, Goodchild, M.F., Parks, B.O. & Steyaert, L.T. (Eds.), Oxford University Press, Oxford, United kingdom, pp. 231-237.

8. Favis-Mortlock, D.T., Boardman, J. & MacMillan, V.J. (2001). The limits of erosion modeling: why we should proceed with care. In: Modeling Landscape Erosion and Evolution, Harmon, R. & Doe, W.W. III (Eds.), Kluwer, New York, USA, pp. 477-516.

9. Ferreira, V.A., Weesies, G.A., Yoder, D.C., Foster, G.R., & Renard, K.G. (1995). The site and condition specific nature of sensitivity analysis. Journal of Soil and Water Conservation, Vol. 50, No. 5, pp. 493-497.

10. Francia, J.R., Duran Zuazo, V.H. & Martinez, A. (2006). Environmental impact from mountainous olive orchards under different soil management systems (SE Spain). Science of the Total Environment, Vol. 358, pp. 46–60.

11. Gago, P., Cabaleiro, C. & Garcìa, J. (2007). Preliminary study of the effect of soil management systems on the adventitious flora of a vineyard in northwestern Spain. Crop Protection, Vol. 26, pp. 584–591.

12. Gomez, J.A., Álvarez, S. & Soriano, M.A. (2009a). Development of a soil degradation assessment tool for organic olive groves in southern Spain. Catena, Vol. 79, pp. 9–17.

13. Gomez, J.A., Battany, M., Renschler, C.S. & Fereres, E. (2003). Evaluating the impact of soil management on soil loss in olive orchards. Soil Use and Management, Vol. 19, pp. 127–134.

14. Gomez, J.A. & Giraldez, J.V. (2007). Soil and water conservation. A European approach through ProTerra projects. Proceedings of the European Congress on Agriculture and the Environment, Sevilla, Spain,

26–28th, 2007.

15. Gomez, J.A., Giraldez, J.V. & Fereres, E. (2005). Water erosion in olive orchards in Andalusia (Southern Spain): a review. Proceedings of the general assembly of the European Geophysical Union. Wien, Austria, April 24–29th.

16. Gomez, J.A., Guzman, M.G., Giraldez, J.V. & Fereres, E. (2009b). The influence of cover crops and tillage on water and sediment yield, and on nutrient, and organic matter losses in an olive orchard on a sandy loam soil. Soil & Tillage Research, Vol. 106, pp. 137–144.

17. Gomez, J.A., Romero, P., Giraldez, J.V. & Fereres, E. (2004). Experimental assessment of runoff and soil erosion in an olive grove on a Vertic soil in Southern Spain as affected by soil management. Soil Use and Management, Vol. 20, pp. 426–431.

18. Gomez, J.A., Sobrinho, T.A., Giraldez, J.V. & Fereres, E. (2009c). Soil management effects on runoff, erosion and soil properties in an olive grove of Southern Spain. Soil & Tillage Research, Vol. 102, pp. 5–13.

19. Helling, J. & Haigh, M.J. (2002). Better land husbandry in Honduras: towards the new paradigm in conserving soil, water and productivity. Land Degradation & Development, Vol. 13, pp. 233–250.

20. Kosmas, C., Danalatos, N., Cammeraat, L.H., Chabart, M., Diamantopoulos, J., Farand, R., Gutiérrez, L., Jacob, A., Marques, H., Martínez-Fernández, J., Mizara, A., Moutakas, N., Nicolau, J.M., Oliveros, C., Pinna, G., Puddu, R., Puigdefábregas, J., Roxo, M., Simao, A., Stamou, G., Tomasi, N., Usai, D. & Vacca, A. (1997). The effect of land use on runoff and soil erosion rates under Mediterranean conditions. Catena, Vol. 29, pp. 45–59.

21. Krause, P., Boyle, D.P. & Base, F. (2005). Comparison of different efficiency criteria for hydrological model assessment. Advances in Geosciences, Vol. 5, pp. 89-97.

22. Legates, D.R. & McCabe, G.J. (1999). Evaluating the use of "goodness of fit" measures in hydrologic and hydroclimatic model validation. Water Resources Research, Vol. 35, pp. 233-241.

23. Licciardello, F., Zema, D.A., Zimbone, S.M. & Bingner, R.L. (2007). Runoff and soil erosion evaluation by the AnnAGNPS model in a small Mediterranean watershed. Transactions of the ASABE, Vol. 50, No. 5, pp. 1585-1593.

24. McCool, D.K., Foster, G.R., Mutchler, C.K. & Meyer, L.D. (1989). Revised slope length factor for the Universal Soil Loss Equation. Transactions of the ASAE, Vol. 32, No. 5, pp. 1571-1576.

25. Monteiro, A. & Moreira, I. (2004): Reduced rates of residual and postemergence herbicides for weed control in vineyards. Weed Research, Vol. 44, pp. 117–128.

26. Montgomery, D.R. (2007). Soil erosion and agricultural sustainability. Proceedings of the National Academy of Sciences, Vol. 104, pp. 13268–13272.

27. Morgan, R.P.C. & Nearing, M.A. (2000). Soil erosion models: present and future. Keynotes of the IIIrd International Congress of the European Society of Soil Conservation. Valencia, Spain, pp. 145-164.

28. Nash, J. E. & Sutcliffe, J. V. (1970). River flow forecasting through conceptual models: Part I. A discussion of principles. Journal of Hydrology, Vol. 10, No. 3, pp. 282-290.

29. Pastor, M., Castro, J., Vega, V. & Humanes, M.D. (1999). Sistemas de manejo del suelo. In: El cultivo del olivo. Barranco, D., Fernandez-Escobar, R. & Rallo, L. (Eds.), Mundi Prensa, Madrid, Spain (in Spanish).

30. Raglione, M., Toscano, P., Angelini, R., Briccoli-Bati, C., Spadoni, M., De Simona, C. & Lorenzini, P. (1999). Olive yield and soil loss in hilly environment of Calabria (Southern Italy). Influence of permanent cover crop and ploughing. Proceedings of the International Meeting on Soils with Mediterranean Type of Climate, July 4th–9th, 1999, University of Barcelona, Barcelona, Spain.

31. Ramos, M.C. & Martìnez-Casasnovas, J.A. (2004). Nutrient losses from a vineyard soil in Northeastern Spain caused by an extraordinary rainfall event. Catena, Vol. 55, pp. 79–90.

32. Renard, K.G., Foster, G.R., Weesies, G.A., McCool, D.K. & Yoder, D.C. (1997). Predicting soil erosion by water: a guide to conservation planning with the revised universal soil loss equation (RUSLE). US Department of Agriculture Agricultural Handbook No. 703. USDA, Washington DC, USA.

33. Renard, K.G. & Ferreira, V.A. (1993). RUSLE model description and database sensitivity. Journal of Environmental Quality, Vol. 22, No. 3, pp. 458-466.

34. Renard K.G., Foster G.R., Weesies G.A. & Porter J.P. (1991). RUSLE Revised Universal Soil Loss Equation. Journal of Soil and Water Conservation, Vol. 46, No. 1, pp. 30-33.

35. Shrestha, S., Babel Mukand, S., Das Gupta, A. & Kazama, F. (2006). Evaluation of annualized agricultural nonpoint source model for a watershed in the Siwalik Hills of Nepal. Environmental Modelling and Software, Vol. 21, No. 7, pp. 961-975.

36. Stone, R.P. & Hilborn, D. (2000). In: Fact Sheet: Universal Soil Loss Equation (USLE). OMAFRA, n.d., available from http://www.omafra. gon.on.ca/english/engineer/facts/00-001.htm.

37. Taguas, E. V., Cuadrado, P., Ayuso, J.L., Yuan, Y., & Perez, R. (2010). Spatial and temporal evaluation of erosion with RUSLE: a case study in an olive orchard microcatchment in Spain. Solid Earth Discussion, Vol. 2, pp. 275–306, 22.03.2010, Available from www.solid-earth-discuss. net/2/275/2010/doi:10.5194/sed-2-275-2010.

38. Tiwari, A.K., Risse, L.M. & Nearing, M.A. (2000). Evaluation of WEPP and its comparison with USLE and RUSLE. Transactions of the ASAE, Vol. 43, pp. 129-1135.

39. Van Liew, M.W. & Garbrecht, J. (2003). Hydrologic simulation of the Little Washita River experimental watershed using SWAT. Journal of the American Water Resources Association, Vol. 39, pp. 413-426.

40. Wischmeier, W. & Smith, D. (1978). Predicting rainfall erosion losses: a guide to conservation planning. US Department of Agriculture Agricultural Handbook No. 537, USDA, Washington DC, USA.

41. Yoder, D.C, Foster, G. R., Weesies, G. A., Renard, K. G., McCool, D. K. & Lown, J. B. (2001). Evaluation of the RUSLE Soil Erosion Model, n.d., available from http://www3.bae.ncsu.edu/Regional-Bulletins/Modeling-Bulletin/rusle-yoder- 01016.html.

42. Zema, D.A., Bingner, R.L., Govers, G., Licciardello, F., Denisi, P. & Zimbone, S.M. (2011). Evaluation of runoff, peak flow and sediment yield for events simulated by the AnnAGNPS model in a Belgian agricultural watershed. Land Degradation and Development (in press). Available on line since 14/12/2010.

Chapter 10

HYDROLOGICAL MODEL TO SIMULATE DAILY FLOW IN A BASIN WITH THE HELP OF A GIS

Vitali Diaz Mercado[1], Khalidou M. Bâ[2], Emmanuelle Quentin[3], Febe Helia Ortiz Madrid[2], AND Lilly Gama[4]

[1]UNESCO-IHE Institute for Water Education, Delft, Netherlands

[2]Centro Interamericano de Recursos del Agua, Universidad Autónoma del Estado de México, Toluca, México

[3]Instituto Nacional de Investigación en Salud Pública, Quito, Ecuador

[4]División Académica de Ciencias Biológicas, Universidad Juárez Autónoma de Tabasco, México

ABSTRACT

Hydrological modeling is an essential tool to evaluate water resources in hydrological basins. The time invested in it depends on the structure of the hydrological model chosen, the amount and quality of information required and the efforts invested in calibration. CEQUEAU is a distributed hydrological model developed at the INRS-ETE, Quebec, Canada. The basin is divided into cells and the rainfall-runoff process is simulated cell by cell until the outlet. Recent advances in geomatics make it possible to develop modules integrated in geographic information systems (GIS) to facilitate the processing of information required by hydrological models. The objective of the present investigation is to implement the CEQUEAU model in Idrisi GIS for the hydrological modeling of basins, thereby reducing information processing time and improving limitations in the original version, such as the number of discretization cells and methods to calculate evapotranspiration. This document presents the results from the implementation of the CEQUEAU model, including evapotranspiration, water levels (in reservoirs, soil and aquifers) and hydrographs. These results show that these new changes provide more hydrology options to the user and with better results.

INTRODUCTION

Surface hydrology has developed considerably as a science primarily because of systematic knowledge of the land phase of the hydrological cycle, its complexity and the difficulty of obtaining exact and detailed meteorological and hydrometric observations for large areas of drainage basins.

Hydrological models have emerged from the need to calculate the magnitudes of the variables involved in the water cycle. A model is useful to solve a significant number of hydrological studies, such as: reconstitution and generation of long series of data, detection of observation errors, forecasting of extreme events, calculation of flows at ungauged sites, operation of reservoirs and conducting environmental studies, among others. With multidisciplinary approaches, hydrological models are useful to simulate water quality, for example, models that simulate transport of pollutants and those that simulate aquifer levels in agricultural areas, among others [1].

Since the 1970s, the development of computing has stimulated the generation of distributed hydrological models (DHM). Investigations using DHM for large areas are based on the hydrological investigation of processes [2]. Thus, macroscale hydrological models are constructed which can be executed repeatedly over large geographic areas.

DHM have been used to evaluate hydrological conditions (runoff, infiltration, aquifer recharge), the state of vegetation (density, quality) and climate change over large regions. In fact, distributed models can be applied to any type of hydrological problem, including forecasting in basins with no instrumentation.

The use of GIS in hydrological modeling has become more widespread over recent decades. For example, in 1998, the Center for Research in Water Resources (CRWR) at the University of Texas created CRWR-PrePro, a pre-processor in ArcView which extracts information from digital spatial data and makes it available for use by the hydrological program HEC-HMS, which calculates flows [3] - [4].

Molnar and Julien [5] modeled runoff in a basin using the CASCA2D distributed model. Fortin et al. [6] - [7] developed the HYDROTEL distributed model for the purpose of its compatibility with GIS and applied it to the Chaudière River basin (Canada). Chávez and Estrada [8] in the Idrisi Mexico Resources Center (CRI-Méx, Spanish acronym), developed an interface between the CEQUEAU model and ArcView GIS which enables extracting the information required to simulate flows.

It is evident that significant developments have occurred in hydrological models over the past three decades. Three primary factors are involved in these developments: 1) technological advances in geographic information systems (GIS); 2) availability of digital elevation models (DEM) used in GIS; and 3) availability of various digital databases (climatological and hydrometric).

This has gradually made it possible to more quickly and accurately obtain the parameters required by hydrological models [9]. The quality of the results depends on the accuracy of the input data and the degree to which the model's structure correctly represents the hydrological process of the problem studied.

OBJECTIVES

The overall objective of this investigation is to implement the CEQUEAU distributed model in the Idrisi geographic information system and apply that model to the study of a hydrological basin to analyze the efficiency and speed of this new tool to simulate flows.

The specific objectives are: implement the CEQUEAU distributed model in a geomatics framework as an additional application of the Idrisi geographic information system; analyze the land use and hydrometeorological information available in a study basin to organize and generate geodatabases; apply the hydrological model implemented in Idrisi to a basin and analyze the results.

MATERIALS AND METHODS

Cequeau Hydrological Model

This model was developed at the National Institute for Water-Scientific Research (formerly INRS-EAU, now INRS-ETE, French acronyms) at the University of Quebec, Canada to reproduce runoff in a basin [10]. This has been used in different countries for the continuous simulation of flows and for hydrological forecasting for reservoir management [11] - [12]. It is a distributed model in which the basin is divided into square elemental surfaces (parcels) and flows are calculated for each one, taking into account the spatial-temporal variations in the physiographic characteristics. The model consists of two parts to simulate vertical and horizontal water flow in each square (Figure 1). The first part is called the production function and the second is the transfer function.

The production function refers to the modeling of vertical water flow (rainfall, evapotranspiration, infiltration, etc.). This function is aimed at obtaining the water volume for each one of the three recipients included in the model: lakes-marshes, soil and aquifer.

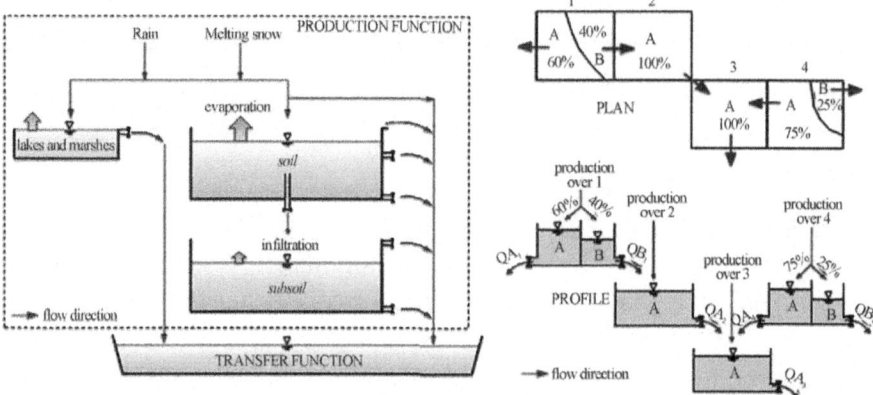

Figure 1. Production function (left) and transfer function (right) for the distributed hydrological model.

The water volume is calculated for each partial element by multiplying the water depth produced in the entire square by the area of the partial element under consideration. The transfer function analyzes the way in which the flow is transferred through the drainage network, taking into account lakes, marshes, dams and bypasses, among other factors. The model examines each parcel for defined time intervals, which can be one day or even one hour.

Evapotranspiration is calculated based on the modified Thornthwaite formula [10]. The calculations are performed according to the time intervals indicated by the user. As shown in the illustration inFigure 1, surface and delayed runoff depend on the water level in the recipients, conceptual draining coefficients and the thresholds of the recipient. This is simulated using equations for the model's parameters which govern the behavior of the water flow in the three recipients (reservoir, soil and aquifer), snow melting, evapotranspiration and channel flow routing.

Two types of input data are required by the model, physiographic and hydrometeorological. The physiographic data are processed for each of the parcels into which the basin is discretized. The use of two-thirds of the hydrometric data is recommended to calibrate the model (estimation of parameters) and the remaining one-third is recommended to validate it.

The model has an application to optimize the model's parameters. The algorithm is based on the Powell method, whose objective function is the Nash coefficient or the correlation coefficient (r). The equations for these calculations are shown by Equations (1) and (2), respectively.

$$NASH = 1 - \frac{\sum_{i=1}^{n}(Qc_i - Qo_i)^2}{\sum_{i=1}^{n}(Qo_i - \overline{Qo})^2}$$

(1)

$$r = \frac{n\sum_{i=1}^{n}(Qo_i Qc_i) - \left(\sum_{i=1}^{n} Qo_i\right)\left(\sum_{i=1}^{n} Qc_i\right)}{\sqrt{n\left(\sum_{i=1}^{n} Qo_i^2\right) - \left(\sum_{i=1}^{n} Qo_i\right)^2} - \sqrt{n\left(\sum_{i=1}^{n} Qc_i^2\right) - \left(\sum_{i=1}^{n} Qc_i\right)^2}}$$

(2)

where the Nash coefficient is dimensionless; Qc_i is the calculated flow on day i (m³/s); Qo_i is the observed flow on day i (m³/s) and; $\overline{Qo}$ is the average of the observed flows n (m³/s). The Nash coefficient ranges from $-\infty$ to 1, where the value 1 corresponds to perfect simulation.

Methodology

To implement the CEQUEAU model in Idrisi, the method proposed by Quentin et al. [13] to develop geomatics modules was followed:

- The hydrological model was conceptualized for its incorporation in a GIS environment. This primarily involves considering the spatial variability of the information (rain, temperature, land use).
- Based on the conceptual model, the geomatic model was developed taking into account the structures and types of operations available in the GIS. Since different geomatics models can be constructed with the same conceptual model, it was necessary to identify the one most useful for the proposed objectives and requirements.
- The model was implemented as a GIS geomatics model, that is, the algorithm was implemented in programming language.
- Lastly, the implemented hydrological model was tested with various applications to correct the algorithm and validate the results.

Implementation of the CEQUEAU Model in Idrisi

The hydrological CEQUEAU model was written in Fortran and the applications in Idrisi were implemented in Delphi frameworks base on Pascal language. To this end, the algorithms for the hydrological model needed to be analyzed in order to facilitate the implementation in Delphi.

The number of squares (parcels) into which a basin can be discretized increased, given that the CEQUEAU hydrological model is limited to a maximum of 1000 squares. Nevertheless, since a basin can now be discretized up to the size of the cell (pixel), the dimensions of the square would be defined by the resolution of the matrix file (size of the pixel). Figure 2 illustrates this improvement, where two discretization schemes of a basin are presented, in a) the first scheme the watershed is divided into 235 squares, in b) the second one, into 1,905,015 squares. The Increased in resolution is 100 times more.

The limit of weather stations increased since previously it was only possible to interpolate information for up to 100 stations, which was not convenient when a larger amount existed. Nevertheless, it is now possible to process information from however many stations are available. Also, estimated meteorological data from radar or satellite can now be used.

The transfer function of the CEQUEAU model takes into account the effect of land use (forest, lakes and marshes) in a constant manner that is, based on only one physiographic file for the entire simulation period. Thus, the forest, lake and marsh areas are assumed to remain unchanged over time. In the CEQUEAU model implemented in Idrisi more than one physiographic file can be used.

The modified Thornthwaite formula [10] is the only method used by the CEQUEAU model to calculate evapotranspiration. Four more criteria were now implemented: Makkink, Turc, Hargreaves-Samani and Penman-Monteith FAO 56. The first three methods mentioned were implemented because they are considered to be more applicable in the different climatic environments in Mexico; the latter was used because it is considered to be the most reliable of the evapotranspiration methods [14].

Proposed Geodatabase

To develop any tool used to perform different applications, the extraction sources for the input information must be defined, since its use is not possible without these data. The present investigation proposes and uses four sources from which input information for the CEQUEAU implemented in the Idrisi GIS can be obtained: CLICOM, BANDAS, USGS and CONABIO.

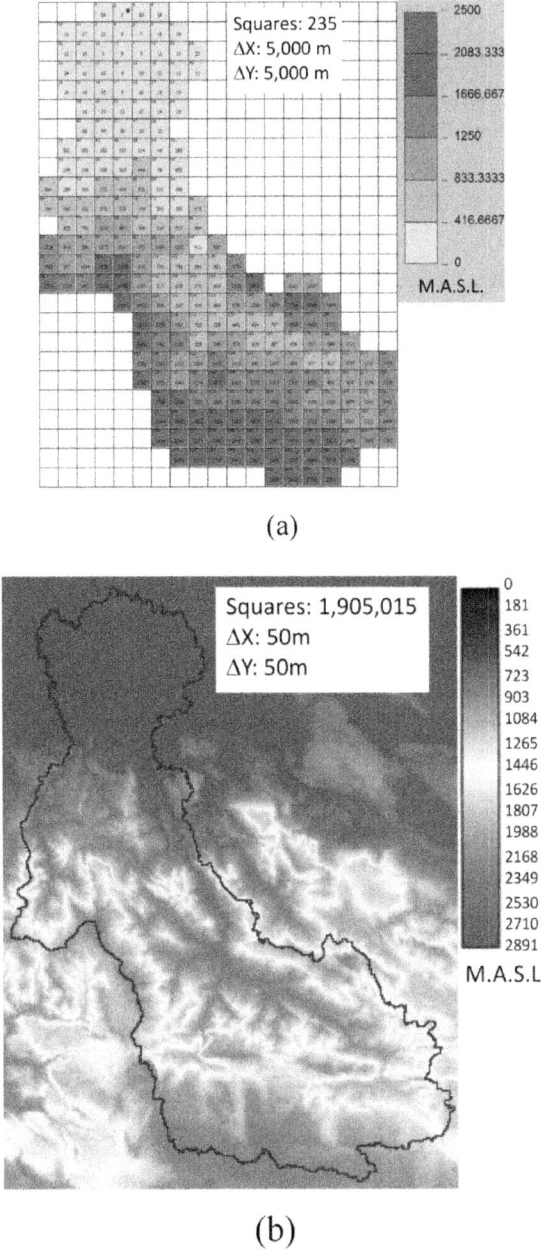

(a)

(b)

Figure 2. Discretization schemes: (a) 235 squares (limit 1000) and (b) 1,905,015 squares.

Meteorological & Hydrometric Data

Precipitation and maximum and minimum daily temperatures were obtained from the CLICOM database, which can be acquired from the National Weather Service (SMN, Spanish acronym). CLICOM is the database from which the ERIC (Spanish acronym) database is obtained (Rapid Extractor of Climate Information, Extractor Rápido de Información Climatológica) which was also used by this study.

The hydrometric data were taken from the National Surface Water Data Bank (BANDAS, Spanish acronym). This database was obtained from the Mexican Institute for Water Technology (IMTA, Spanish acronym).

Digital Elevation Model

The topography of the study area is represented by an image from the Digital Elevation Model (DEM) generated by the US Geological Survey (USGS) based on radar images. The information is raster (matricial) with a resolution of 90 m × 90 m. The DEM image was downloaded from the USGS site (http://earthexplorer.usgs.gov/), introducing the geographic coordinates of the polygon representing the study.

Land Use

The land use image for Mexico was obtained from the National Commission for Knowledge and Use of Biodiversity (CONABIO, Spanish acronym) and are vector images at a scale of 1:250,000. The commission developed them by digitalizing topographic maps of the Mexican Republic at that scale. CONABIO developed an interactive download site from which the information can be obtained (http://www.conabio.gob.mx/informacion/gis/).

Calibration of the Hydrological Model

The similarity between the observed and measured (Vo) values and those that are simulated (Vc) using the hydrological model can be measured or estimated using an objective function or numerical criterion. In general, flow is the variable of greatest interest to calculate using a hydrological model. Nevertheless, depending on the results and the conception of the model, other comparisons can be performed (for example, soil moisture content and evapotranspiration, among others). Table 1 presents four of the objective functions used as evaluation criteria by this work. It also shows the optimal values indicated by a perfect simulation.

RESULTS FROM THE IMPLEMENTATION

After implementing the CEQUEAU hydrological model in Idrisi, tests were performed to verify its accuracy. These consisted of comparing the results obtained using the developped application with those from the original model for the same set of input data. Figures 3-6 present evapotranspiration, water levels in the soil (HS) & aquifer (HN) and flow, respectively, calculated with the CEQUEAU model and the model implemented in Idrisi. In all cases, determination coefficients (R^2) and Nash coefficients very close to one were obtained, confirming the correct implementation of the hydrological model.

APPLICATION

The CEQUEAU hydrological model implemented in Idrisi was used to simulate daily flows for the La Sierra River basin (Figure 7), which begins in the central high plains of Chiapas and is 227 km long with a mean slope of 0.96%, anditis located between coordinates 97°40'50"N, 16°40'47"W and 92°3'22"N, 17°51'38"W. This basin is long and exorheic (5 order) with high reliefs upstream and flat at the outlet. Its area is 4800 km² and its perimeter measures 686 km. It is located in hydrological region No.30 Grijalva-Usumacinta where the annual regional precipitation is 4000 mm.

Table 1. Most commonly used objective functions to quantify the modeling error.

Determination Coefficient	Relative Mean Square Error	Nash Coefficient	Coefficient of Variation Error
$R^2 = \left[\dfrac{\sum (Vo_i - \bar{Vo})(Vc_i - \bar{Vc})}{\sqrt{\sum (Vo_i - \bar{Vo})^2 \sum (Vc_i - \bar{Vc})^2}} \right]^2$	$ERCM = \dfrac{1}{n}\sum_{i=1}^{n}\left[\dfrac{Vo_i - Vc_i}{Vo_i}\right]^2$	$NASH = 1 - \dfrac{\sum (Vc_i - Vo_i)^2}{\sum (Vo_i - \bar{Vo})^2}$	$CVE = \dfrac{\sqrt{\frac{1}{n-1}\sum\left[(Vo_i - Vc_i) - \frac{1}{n}\sum (Vo_i - Vc_i)\right]}}{\frac{1}{n}\sum (Vo_i - Vc_i)}$
Optimal Value: $R^2 = 1$	ERCM = 0	NASH = 1	CVE = 0
Range: [0, 1]	[0, ¥)	(−¥, 1]	[0, ¥)

Vo_1: nth observed value, Vc_1: nth calculated value, $\bar{Vo}$: Average n values observed, $\bar{Vc}$: Average n values calculated.

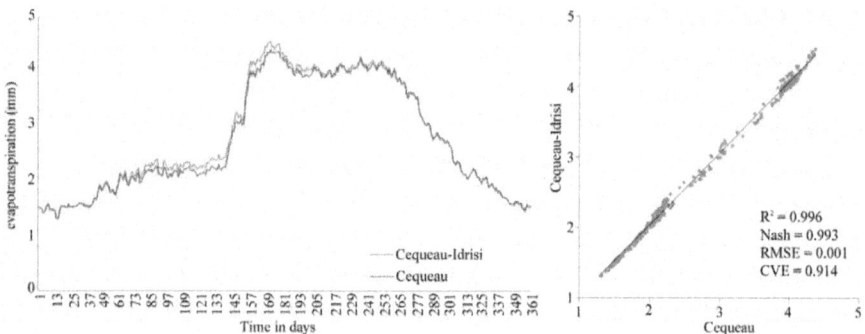

Figure 3. Comparison between inter-annual evapotranspiration (mm) calculated with the CEQUEAU model and the model implemented in Idrisi.

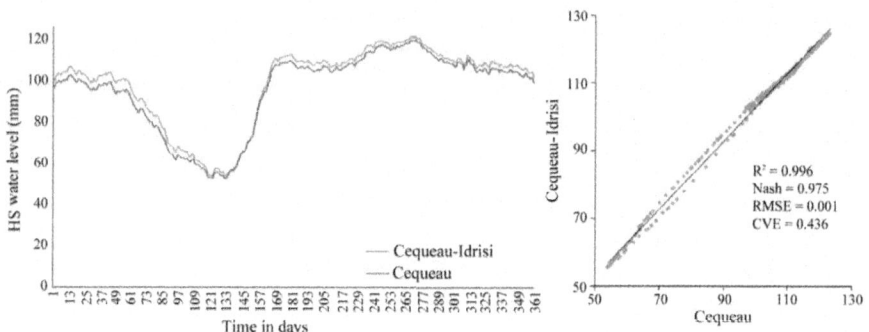

Figure 4. Comparison between inter-annual HS water level (mm) calculated with the CEQUEAU model and the model implemented in Idrisi.

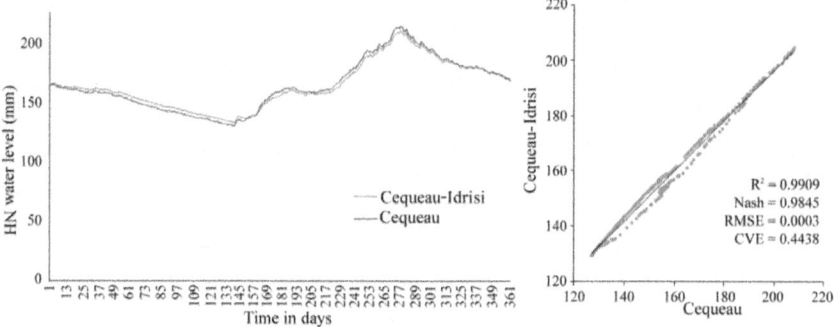

Figure 5. Comparison between inter-annual HN water level (mm) calculated with the CEQUEAU model and the model implemented in Idrisi.

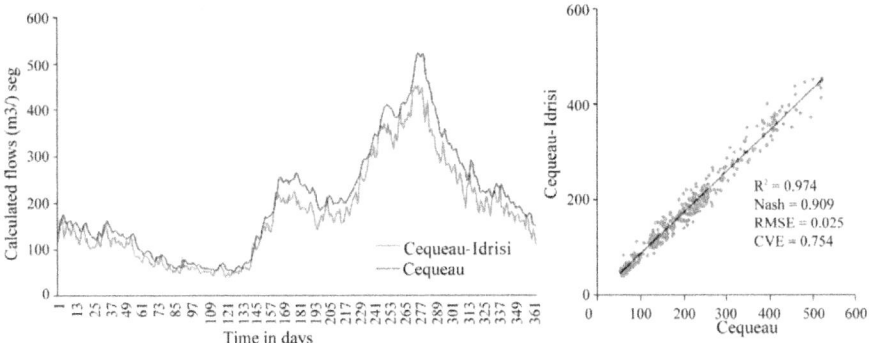

Figure 6. Comparison between inter-annual flows (mm) calculated with the CE-QUEAU model and the model implemented in Idrisi.

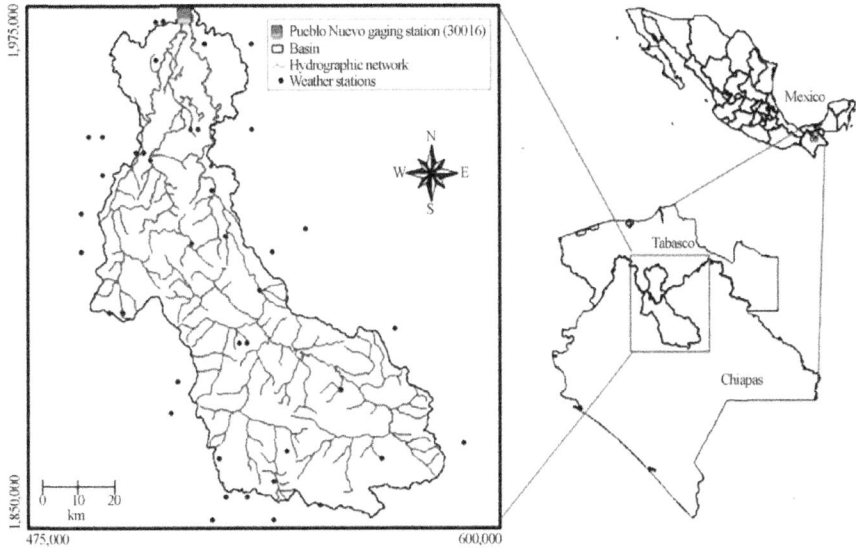

Figure 7. La Sierra River basin Located in Hydrological Region No. 30, Grijalva-Usumacinta, between the States of Chiapas and Tabasco, Mexico.

Figure 8(a) shows a compound image from the DEM of the study area for UTM zone 15 with a resolution of 50 m × 50 m. The highest elevations is around 2800 M.A.S.L. Figure 8(b) presents the land use image in matrix format (raster), reclassified according to the land uses included in the CEQUEAU hydrological model: 1) water bodies, 2) forests, 3) marshes and 4) others.

RESULTS FROM THE APPLICATION

The results from the simulation of inter-annual daily flows and monthly mean flows using the Thornthwaite method to calculate reference evapotranspiration for the La Sierra River basin are shown in Figures 9-10, respectively. The result of the simulation was satisfactory since for the case of inter-annual daily flows a Nash coefficient of 0.88 was obtained for the entire period and for monthly mean flows a Nash coefficient of 0.89 was obtained. The results of the simulation of the inter-annual daily flows and monthly mean flows using the Penman- Monteith FAO 56 method to calculate evapotranspiration are shown in Figures 11-12, respectively. The simulation was satisfactory since for the case of inter-annual daily flows a Nash coefficient of 0.83 was obtained for the entire period and for monthly mean flows a Nash coefficient of 0.84 was obtained. Although these results are lower than those from the simulation using the Thornthwaite method, the model was calibrated only evapotranspiration method and not for the Penman-Monteith FAO 56. In addition, the latter has been shown to be more applicable than the Thornthwaite method for different climates and latitudes [15] and, therefore, the results will improve with better calibration. The list comparing results between daily recorded and calculated flows is shown in Table 2 for the simulation for 1968 to 1999 (32 years).

CONCLUSIONS

The CEQUEAU hydrological model was implemented in the Idrisi GIS software, taking into account the need to use information from all the weather stations with data available, discretization of the basin into more than 1000 squares if necessary, use of other evapotranspiration methods and the analysis of the effect of the physiographic variability.

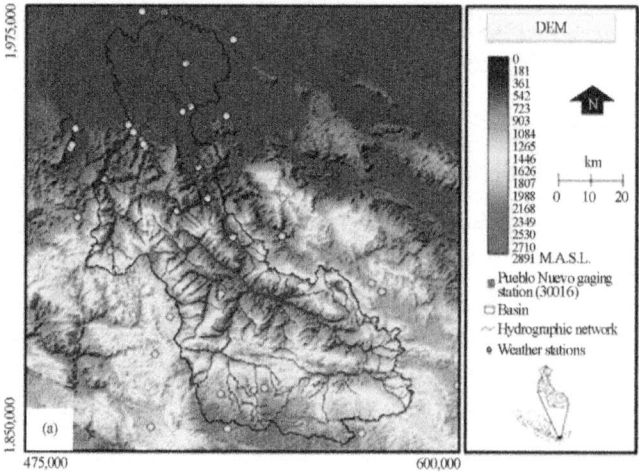

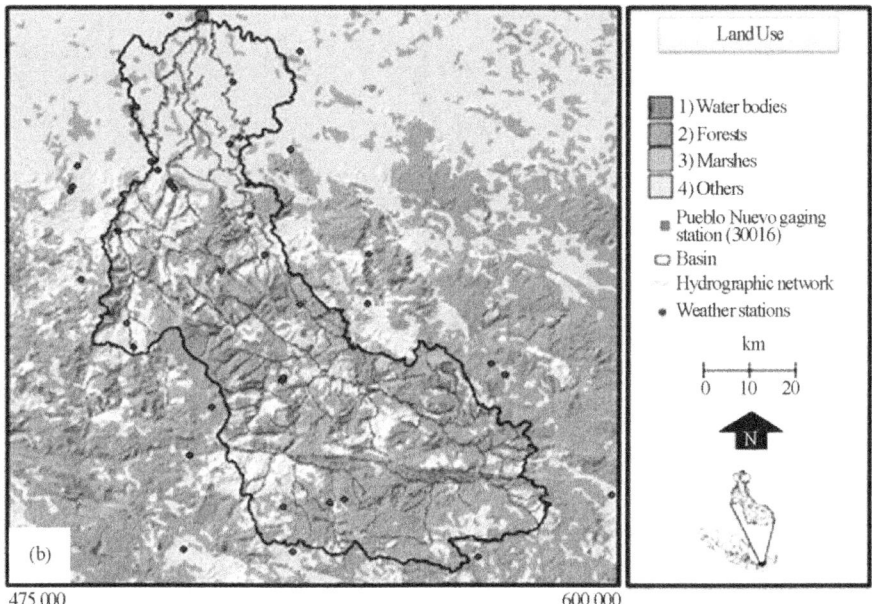

Figure 8. (a) Compound image of the DEM, basin and hydrographic network for the La Sierra River; (b) land use image for the La Sierra River basin.

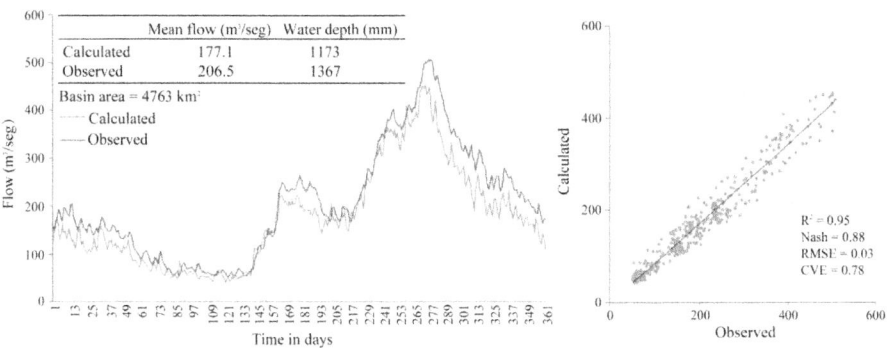

Figure 9. Daily inter-annual calculated and observed flows (m³/sec) at the Pueblo Nuevo station (30016) using the imple- mented CEQUEAU hydrological model, Thornthwaite method.

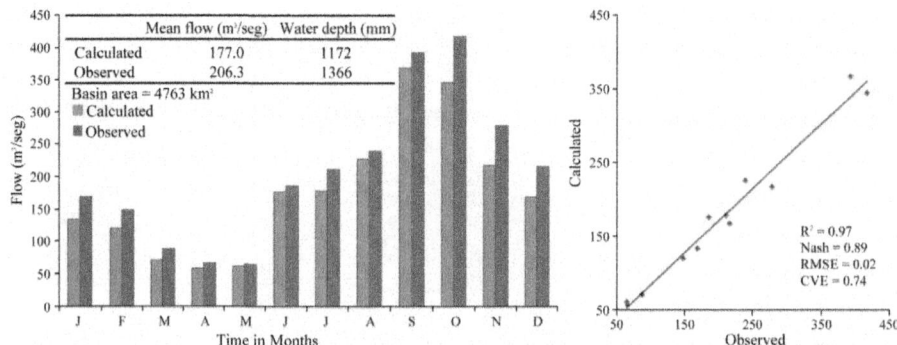

Figure 10. Calculated and observed monthly mean flows (m³/sec) at the Pueblo Nuevo station (30016) using the implemented CEQUEAU hydrological model, Thornthwaite method.

To calculate evapotranspiration, it was possible to implement several calculation methods with a minimal amount of information needed for the analysis, such as maximum and minimum temperatures and the geographic location of the study area. In addition to performing the Thornthwaite method which is used with the CEQUEAU hydrological model, methods that better adapt to Mexico's climatic characteristics were selected: Penman-Mon- teith FAO 56, Hargreaves-Samani, Turc and Makkink.

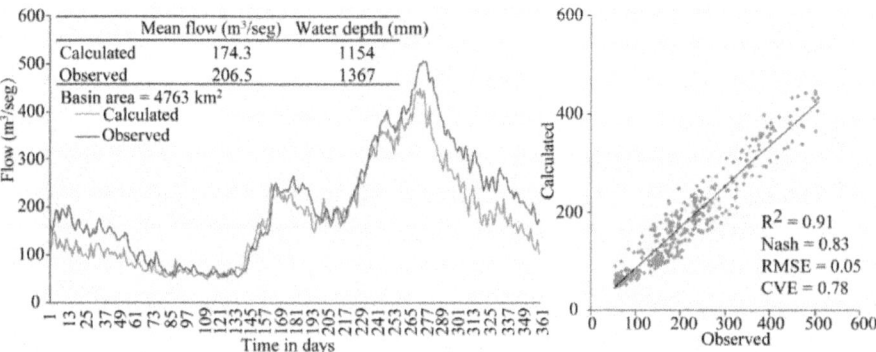

Figure 11. Calculated and observed inter-annual daily flows (m³/sec) at the Pueblo Nuevo station (30016) using the implemented CEQUEAU hydrological model, Penman-Monteith FAO 56.

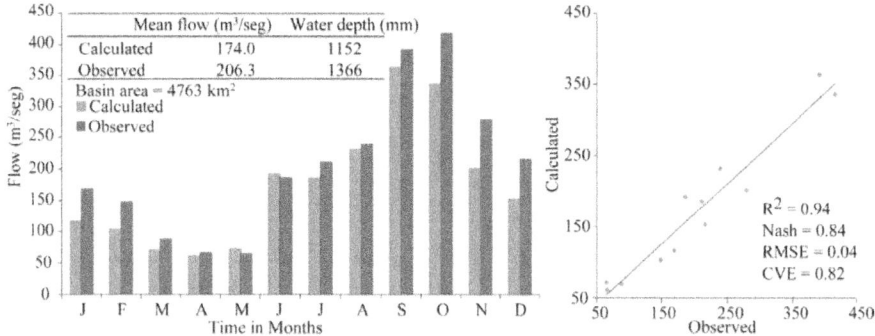

Figure 12. Calculated and observed monthly mean flows (m³/sec) at the Pueblo Nuevo station (30016) using the implemented CEQUEAU hydrological model, Penman-Monteith FAO 56.

Table 2. Numerical criteria to evaluate the simulations of daily flows based on different scenarios.

Year	Thornthwaite				Penman-Monteith FAO 56			
	R^2	Nash	RMSE*	CVE**	R^2	Nash	RMSE*	CVE**
1968	0.88	0.84	0.04	1.26	0.81	0.76	0.07	0.88
1969	0.98	0.95	0.04	1.04	0.93	0.93	0.14	0.69
1970	0.93	0.92	0.03	1.01	0.9	0.9	0.05	0.85
1971	0.91	0.9	0.03	1.05	0.91	0.91	0.03	1.03
1972	0.89	0.88	0.02	1.24	0.76	0.76	0.05	0.93
1973	0.92	0.85	0.04	1.4	0.9	0.84	0.07	1.15
1974	0.95	0.9	0.03	0.9	0.93	0.9	0.05	1.1
1975	0.98	0.98	0.04	0.98	0.96	0.96	0.12	0.75
1976	0.95	0.94	0.02	0.97	0.9	0.88	0.04	0.7
1977	0.94	0.91	0.03	1	0.87	0.84	0.03	0.83
1978	0.95	0.94	0.01	0.84	0.96	0.93	0.02	0.87
1979	0.97	0.86	0.05	0.79	0.93	0.84	0.09	0.8
1980	0.97	0.93	0.03	0.65	0.93	0.9	0.04	0.68
1981	0.94	0.91	0.03	0.97	0.94	0.91	0.02	1.01
1982	0.94	0.91	0.04	0.66	0.97	0.91	0.03	0.92
1983	0.95	0.95	0.02	0.92	0.91	0.9	0.03	0.94
1984	0.98	0.94	0.02	0.85	0.97	0.9	0.05	0.68
1985	0.82	0.71	0.06	0.65	0.73	0.5	0.09	0.54

1986	0.69	0.36	0.13	0.75	0.6	0.2	0.12	0.89
1987	0.58	0.46	0.21	0.77	0.48	0.32	0.28	0.81
1988	0.95	0.95	0.04	0.84	0.93	0.91	0.07	1.02
1989	0.96	0.78	0.09	0.94	0.94	0.69	0.09	1.05
1990	0.94	0.71	0.09	0.84	0.9	0.57	0.1	0.94
1991	0.86	0.63	0.08	1	0.74	0.46	0.1	1.01
1992	0.86	0.71	0.1	0.62	0.85	0.63	0.11	0.68
1993	0.84	0.7	0.08	0.81	0.82	0.6	0.11	0.8
1994	0.73	0.59	0.08	1	0.61	0.34	0.11	1.17
1995	0.91	0.81	0.06	1.04	0.9	0.7	0.1	0.96
1996	0.86	0.24	0.13	0.75	0.78	−0.02	0.17	0.78
1997	0.77	0.26	0.17	0.91	0.72	0.11	0.18	0.95
1998	0.71	0.69	0.47	0.94	0.53	0.51	0.68	1.02
1999	0.96	0.88	0.08	0.81	0.93	0.79	0.1	0.82

*RMSE: relative mean square error; **CVE: coefficient of variation error.

The availability of spatially distributed information, hydrometeorological data and the radar-generated Digital Elevation Model (DEM) helped to create the geodatabase required by the study, which enabled developing and applying the CEQUEAU hydrological model implemented in Idrisi.

The main objective was met, which was to implement the CEQUEAU distributed model in Idrisi and apply this model to the study of a hydrological basin under different scenarios. In addition, the efficiency of this new tool to simulate flows was analyzed and satisfactory results were obtained (Nash coefficients of 0.83 to 0.88 for inter-annual daily flows).

ACKNOWLEDGEMENTS

The authors thank the National Council on Science and Technology (Consejo Nacional de Ciencia y Tecnología (CONACyT, Spanish acronym)) for the scholarship to graduate studies. In addition, this work was financed by the CONACyT research project 90637 and UEAM projects 3459/2013CH and 2752/2009, the latter provided by the Juarez Autonomous University of Tabasco. The authors would like to thank the anonymous reviewers for their constructive comments, which helped to improve the manuscript.

REFERENCES

1. Ba, K.M., Díaz-Delgado, C. and Rodríguez-Osorio, V. (2001) Simulación de caudales de los ríos Amacuzac y San Jerónimo en el Estado de Mexico, Mexico. Ingeniería Hidráulicaen México (IHM), 16, 117-126.

2. Venneker, R.G. and Bruijinzeel, L.A. (1997) The IHE-VUA Cathment Research and Modelling lnifiafive (CRMI). The IHE-VUA Catchment Research and Modelling Initiative CRMI-RN-001. IIHHEE, Delfl, Vrije Universiteit Amsterdam, The Netherlands.

3. Olivera, F., Reed, S. and Maidment, D.R. (1998) HEC-PrePro v. 2.0: An ArcView Pre-Processor for HEC's Hydrologic Modeling System. 1998 ESRI User's Conference, 25-31 July 1998, San Diego.

4. Olivera, F. and Maidment, D.R. (1999) GIS Tools for HMS Modeling Support. 1999 ESRI User's Conference, Sandiego.

5. Molnar, D.K. and Julien, P.Y. (2000) Grid-Size Effects on Surface Runoff Modeling. Journal of Hydrologic Engineering, 5, 8-16. http://dx.doi.org/10.1061/(ASCE)1084-0699(2000)5:1(8)

6. Fortin, J.P., Turcotte, R., Massicotte, S., Moussa, R., Fitzback, J. and Villeneuve, J.P. (2001) A Distributed Watershed Model Compatible with Remote Sensing and GIS Data. I: Description of Model. Journal of Hydrologic Engineering, 6, 91-99.http://dx.doi.org/10.1061/(ASCE)1084-0699(2001)6:2(91)

7. Fortin, J.P., Turcotte, R., Massicotte, S., Moussa, R., Fitzback, J. and Villeneuve, J.P. (2001) A Distributed Watershed Model Compatible with Remote Sensing and GIS Data II: Application to Chaudiere Watershed. Journal of Hydrologic Engineering, 6, 100-108.http://dx.doi.org/10.1061/(ASCE)1084-0699(2001)6:2(100)

8. Chávez, M.I. and Estrada, B. (2005) Programación de una interfaz entre el modelo hidrológico CEQUEAU y el SIG ArcView. Bacherlors Thesis, Universidad Autónoma del Estado de México, Facultad de Ingeniería, Toluca.

9. He, C. and Croley II, T.E. (2007) Application of a Distributed Large Basin Runoff Model in the Great Lakes Basin. Control Engineering Practice, 15, 1001-1011.http://dx.doi.org/10.1016/j.conengprac.2007.01.011

10. Morin, G. and Paquet, P. (2007) Modèle Hydrologique CEQUEAU. Rapport de Recherche no R000926, INRS-ETE.

11. Ayadi, M. and Bargaoui, Z. (1998) Modélisation des écoulements de l'oued Miliane par le modèle CEQUeau. Journal des Sciences Hydrologiques, 43, 741-758.http://dx.doi.org/10.1080/02626669809492170

12. Eleuch, S., Carsteanu, A.A., Ba, K., Magagi, R., Goita, K. and Díaz-Delgado, C. (2010) Validation and Use of Rainfall Radar Data to Simulate Water Flows in the Río Escondido Basin. Stochastic Environmental Research & Risk Assessment, 24, 559-565.http://dx.doi.org/10.1007/s00477-009-0336-9

13. Quentin, E., Díaz-Delgado, C., Gómez-Albores, M.A., Manzano-Solís, L.R. and Franco-Plata, R. (2007) Desarrollo geomático para la gestión integrada del agua. XI Conferencia Iberoamericana de Sistemas de Información Geográfica (XI CONFIBSIG), 21pp.

14. Campos-Aranda, D.F. (2005) Estimación empírica de la ETP en la República Mexicana. Ingeniería Hidráulica en México (IHM), 20, 99-110.

15. Allen, R., Pereira, L., Dirk, R. and Smith, M. (2006) Evapotranspiración del cultivo. Guía para la determinación de los requerimientos de agua para los cultivos. FAO. Roma. ftp://ftp.fao.org/agl/aglw/docs/idp56s.pdf

Chapter 11

IMPACTS OF FOREST FIRES AND CLIMATE VARIABILITY ON THE HYDROLOGY OF AN ALPINE MEDIUM SIZED CATCHMENT IN THE CANADIAN ROCKY MOUNTAINS

Johanna Springer[1], Ralf Ludwig[1], and Stefan W. Kienzle[2,3]

[1]Department of Geography, Ludwig-Maximilians-Universitaet Muenchen (LMU), Luisenstr. 37, D-80333 Munich, Germany

[2]Department of Geography, University of Lethbridge, Alberta Water and Environmental Science Building, 4401 University Drive, Lethbridge, AB T1K-3M4, Canada

[3]Applied Behavioral Ecology and Ecosystems Research Unit, University of South Africa, PO Box 392, Florida, 1710 Pretoria, South Africa

ABSTRACT

This study investigates the hydrology of Castle River in the southern Canadian Rocky Mountains. Temperature and precipitation data are analyzed regarding a climate trend between 1960 and 2010 and a general warming is identified. Observed streamflow has been declining in reaction to a decreasing snow cover and increasing evapotranspiration. To simulate the hydrological processes in the watershed, the physically based hydrological model WaSiM (Water Balance Simulation Model) is applied. Calibration and validation provide very accurate results and also the observed declining runoff trend can be reproduced with a slightly differing inclination. Besides climate change induced runoff variations, the impact of a vast wildfire in 2003 is analyzed. To determine burned areas a remote sensing method of differenced burn ratios is applied using Landsat data. The results show good agreement compared to observed fire perimeter areas. The impacts of the wildfires are evident in observed runoff data. They also result in a distinct decrease in model efficiency if not considered via an adapted model parameterization, taking into account the modified land cover characteristics for the burned area. Results in this study reveal (i) the necessity to establish specific land cover classes for burned areas; (ii) the relevance of climate and land cover change on the hydrological response of the Castle River watershed; and (iii) the sensitivity of the hydrological model to accurately simulate the hydrological behavior under varying boundary conditions. By these means, the presented methodological approach is considered robust to implement a scenario simulations framework for projecting the impacts of future climate and land cover change in the vulnerable region of Alberta's Rocky Mountains.

INTRODUCTION

The southern Rocky Mountains in Alberta, Canada, are an important region for the generation of freshwater runoff especially due to their contribution of melt water in spring months [1,2]. Climate change affects the watersheds in the southern Rocky Mountains with an increase in air temperatures, especially during winter months, and thus a decrease in snow cover. As a consequence, spring melt runoff has been declining [3]. Precipitation changes are more ambiguous in pattern and amplitude, but also contribute to changes in runoff variability. The impact of climate change on southern Albertan watersheds has been investigated by several authors [1,2,3,4,5]. The trend in runoff due to climate variability in the Castle River watershed was analyzed by Rood [5] and Byrne [1]. Land cover changes can also impact the watershed's runoff regime and discharge amounts. Wildfires are an important natural occurrence on the eastern slopes of Alberta [6] and have always played an integral role in the functioning of ecosystems in Canadian forests [7]. However, over the past three decades there has been a distinct trend in increase of areas burned by wildfire [8]. Flannigan *et al.* [9] examined the effect of climate change on wildfires and emphasize that fire projection models combined with climate change predictions indicate not only an increase in the area burned but also in fire season length, the intensity of fires and thus in burn severity [7]. Pierson, *et al.* [10] investigated the impact of wildfires on watershed hydrology and detected a reduced infiltration and an increased erosion in a steep rangeland study side. According to Silins [11], after the 2003 Lost Creek fire in southern Alberta, important changes in the hydrology of the affected rivers could be measured; mean annual flows increased due to reduced actual evapotranspiration, as did peak discharges and sediment concentration.

To analyze changes in discharge patterns due to environmental change, hydrological modeling is broadly applied. Currently, special focus in environmental modeling is set on the investigation of the impacts of climate variability, land cover changes or human induced land degradation [12]. To provide input data for hydrological modeling or to detect changes in land cover, satellite remote sensing is a recommended and widely used method.

Landsat data are available at high spatial and temporal resolution and deliver information in a broad range of wavelengths especially suited to distinguish different land cover [13]. Thus, different types of patterns in land cover can be detected as well as variations when applying a multi-temporal change detection method. Particularly changes of vegetation characteristics can be distinguished by using infrared bands [13].

In this study, the physically based Water Balance Simulation Model (WaSiM) is applied to simulate the hydrological processes in the Castle watershed. The effect of the fire on the entire Castle River watershed has neither been investigated nor modeled. Thus, this study provides an analysis whether the Castle River shows a reaction in streamflow after the fire and if the model is able to simulate runoff behavior with a changed land cover after wildfires.

The study intends answer to the following questions: How has the hydrology of Castle River been changing during the last decades and what are the reasons for that? Is the hydrological model WaSiM able to represent the hydrological conditions? How severe was the wildfire in 2003 and can the effects on hydrology be observed and simulated with an adjusted parameterization?

The objectives of this study are to (i) analyze characteristics and trends of measured climate and streamflow data; (ii) calibrate and validate the hydrological model WaSiM for the watershed; (iii) map burn severity from satellite imagery for a severe wildfire in 2003; and (iv) simulate the impact of climate variability and forest fires on runoff behavior in the Castle River watershed.

STUDY AREA

The Castle River watershed is situated at the eastern slopes of the Rocky Mountains in southwestern Alberta (Figure 1). It is located between the Waterton Lakes National Park in the south, the Crowsnest River watershed in the north and the border to British Columbia in the west. In the east, the Castle River drains into the plains and provides water for a wide variety of purposes, in particular for irrigated agriculture [14].

The gauge is situated near the town of Beaver Mines (at 49°2919 North and 114°839 West) and defines the watershed for this study with a size of 820.70 km^2 [15]. The watershed has an elevation difference of more than 1600 m as the gauge is at 1187 m and the Castle Mountain has an altitude of 2766 m [16]. The mean annual streamflow of the Castle River at the Beaver Mines gauge is 15.57 m^3·s^{-1} (runoff data from 1960 to present available), with an explicit maximum in May and June and minimum runoff in the winter months.

The peak of the hydrograph in early summer reflects the convergence of the period of peak melt and the most rainy period of the year in June [11].

The Castle basin has frequently been affected by wildfires [11]. The Lost Creek wildfire in summer 2003 was one of the most severe in the upper eastern slopes in many decades and burned 21,000 ha in both the headwater regions of Crowsnest and northern Castle River [17], where it spread in the catchment areas of Lynx Creek and Lost Creek. The 2003 fire perimeter is indicated in Figure 1. The fire burned in all structural forest strata as a so-called crown fire and expanded quite quickly so that the landscape was changed significantly. As the Lost Creek fire took place along two of the tributaries of the Oldman River, *i.e.*, the Castle and the Crowsnest Rivers, local and regional water-based resources were impacted through these forest disturbances [11].

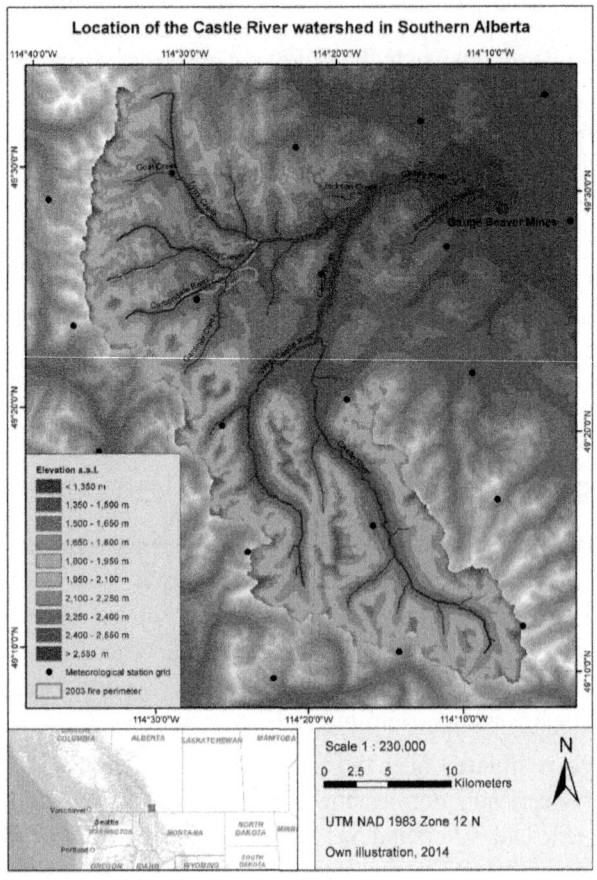

Figure 1. Location of the Castle River watershed in the Southern Rocky Mountains, indicating the burned area of the 2003 Lost Creek fire and the applied meteorological station grid available from AAFC [18].

DATA AND METHODS

Climate and Streamflow Observations and Trends

Historically, meteorological monitoring in the eastern slopes of the Rocky Mountains is extremely sparse [4]. The only available data in the area are interpolated daily minimum and maximum temperatures (in °C) as well as daily precipitation sums (in mm) from AAFC [18] from 1950 to 2010. This data is organized in a regular station grid with a spacing of 10 km, which contains interpolated point estimates. It is derived from an Anusplin interpolation method of Environment Canada climate stations south of 60°N in which a thin-plate smoothing spline surface fitting method was implemented using longitude, latitude and elevation [19]. The situation of the station grid points is represented in Figure 1. Hence, the meteorological data used in this study represent variations from real measured values but they demonstrate general patterns and trends. Climate features are analyzed over a time period of 61 years to investigate the data regarding a climate change trend. To analyze mean temperature characteristics in the region of Castle River watershed, a mean of all 26 stations values is calculated for every day. The same procedure is applied to the precipitation data. Additionally, to analyze spatial variability in the watershed, spatially distributed analyses are accomplished for which the individual station values are considered.

Stream discharge data from Castle River are available at the gauge near the town of Beaver Mines, operated by Environment Canada [15]. The location of the gauge is indicated in Figure 1. Daily runoff values from 1960 to 2010 are analyzed regarding runoff regime, quantitative or temporal changes and trends.

Hydrological Modeling with WaSiM ETH

The hydrological modeling in this study is executed by the mainly physically based, distributed and deterministic hydrological model WaSiM (Water Balance Simulation Model) [20]. It runs in continuous time steps (one day in this study) and the spatial resolution is a raster of constant grid cell size, which is 100 m in this study. WaSiM allows short-term flood event runoff simulations as well as long-term water balance applications [21]. According to Beckers, *et al.* [22], it delivers very high model functionality and complexity for forest management and climate change applications and is, therefore, applied in this study. A detailed description of all modules, algorithms and the functioning of the model is specified in the WaSiM Model Description by Schulla [20]. In this application the model version 9.1.0 is applied, which integrates the Richard-equation.

WaSiM Input Data

WaSiM requires meteorological and spatially distributed input data. The main meteorological input data for an optimal configuration are spatially explicit temperature (°C), precipitation (mm/time step), wind speed (m·s^{-1}), relative humidity (1/1) or vapor pressure (mbar) and either sunshine duration (1/1) or global radiation (Wh·m^{-2}) for each time step. The data are acquired from several meteorological station points distributed in and around the considered watershed so that an interpolation in WaSiM delivers a comprehensive meteorological information.

As AAFC [18] provides minimum and maximum temperature and precipitation data in daily resolution [19] but WaSiM requires daily mean temperature, the mean values are calculated with the equation (Tmax + Tmin)/2, according to Klein Tank, *et al.* [23].

For the study area wind speed, global radiation and relative humidity are only available as monthly means from a spline interpolation of climate normals from the National Climate Data Information Archive [24]. Considering wind speed, in this study the available monthly mean values are used instead of daily ones. For relative humidity and global radiation an estimation of daily data based on air temperature and precipitation data is applied using equations after Thornton, *et al.* [25] and Bristow and Campbell [26]. The meteorological data are thus supplemented by information about the geographical situation and altitude of each station.

Temporal constant geographical data in a grid format are required for a digital elevation model (DEM), land use and soil types. The DEM [27] contributes information about the elevation, the slope, the exposition and the curvature of the surface. Additionally, hydrologic information is derived from the DEM using the program Topographic Analysis (TANALYS) from Schulla [20]. Land cover information in 100 m spatial resolution for the Castle watershed are provided by the GeoBase Land Cover Product [28].

However, an official soil map with soil texture for the Castle River area is not available. WaSiM requires information about the soils in order to calculate infiltration and evaporation processes which depend essentially on the soil texture [20]. Thus, a soil texture map is generated, mainly based on the land cover information and the predominant soil types in different vegetated areas, according to the Canadian Society of Soil Science [29]. In the application of the model, the different soil texture classes obtain specific attributes in several soil layers like the hydraulic conductivity, the saturated and residual water content and other parameters that depend on the grain-size composition of the soil [20].

Calibration and Validation

Calibration of WaSiM is executed gradually. Sensitive model parameters to be calibrated in the Richards-Approach version of WaSiM are the soil model parameters of the unsaturated zone, especially the scaling parameter for interflow and the recession parameter for base discharge for saturated hydraulic conductivity with increasing soil depth (controls peak runoff and base flow) [12].

The calibration period is defined from 1 November 1960 until 31 October 1970, so it equals 10 hydrological years. A comparison of the modeled and observed data sets is conducted and the hydrological goodness of fit criteria NSE and R^2 are calculated in daily and monthly resolution [30,31,32]. Furthermore, the mean annual runoff (MQ), the monthly mean runoff (MMQ), as well as mean high (MHQ) and mean low flow (MLQ) values are compared. Calibration steps are iteratively conducted until a maximum in model efficiency is reached.

An elevation dependent correction for precipitation data is applied separately for rain and snow to derive altitudinal gradients. A threshold temperature between rain and snow have to be chosen for which the value 1 °C showed the best model performance, which is also the threshold temperature for rain in the snow model. The method used is a combination of inverse distance weighting and an elevation dependent regression with internal processing making use of all available meteorological observations. The stomatal resistance parameter was calibrated and needed to be increased to obtain realistic values for actual evapotranspiration. During this calibration process the model efficiency is increased until a stable parameter set is gained.

In order to validate the established model the calibrated parameter set is applied for the following 30 years. Validation runs are executed for each of the decades from 1970 to 1980, from 1980 to 1990, from 1990 to 1999 and from 2000 to 2010.

Remote Sensing of Forest Fires

Spatial wildfire data from ESRD Alberta [33] delivers perimeters of historical big fire events in Alberta. The shapefile polygons are derived from aerial photos and show, beside other fire events, the expansion of the wildfire in 2003, however they lack information on burn severity, which is critical to assess the impact of fires on hydrological variables.

Data

To determine the grade of damage of the burned land cover, infrared Landsat satellite data are applied for vegetation change detection as these wavelengths react sensitive for water content and chlorophyll [34]. Landsat Thematic Mapper 5 (TM5) and Enhanced Thematic Mapper+ (ETM+) data, provided in GeoTIFF format, are used at a spatial resolution of 30 m. Pre- and post-fire images from the 2003 wildfire are chosen according to smallest cloud cover and seasonal similarity to ensure analogous phenology and minimized sun angle effects. Pre-fire imagery is taken from 2002, post-fire imagery stems from shortly after the wildfire was extinguished.

Normalized Burn Ratio and Change Detection Algorithm

The Normalized Burn Ratio (NBR) has been developed for the detection of fire scars [35]; it is explained in detail by Miller and Thode [13] and Miller, *et al.* [36]. The NBR is computed from Landsat bands 4 and 7, the near and short wave infrared bands [36]. Analyzing pixels from pre- and post-fire images, both bands demonstrate considerable changes especially in forested regions. The reflectance in band 4 decreases because of a loss of the photosynthetic activity while the reflectance of band 7 increases due to less water absorption.

Miller and Thode [13] emphasize that vegetated pre-fire areas have different NBR values as the value refers to the density in vegetation cover with its chlorophyll and water content. Hence, to generate comparable fire severity values, a change detection method (differenced Normalized Burn Ratio (dNBR)) is applied in which each post-fire NBR is subtracted from the pre-fire NBR [36]. Low dNBR values indicate unchanged landscape or low fire severity, high dNBR values represent severe damage to vegetation through the fire [13].

The dNBR is meant to normalize the results to pre-fire vegetation cover. However, in several studies the relationship between measured severity values and calculated dNBR has been tenuous [37,38]. Miller and Thode [13] finally present a relative version of the dNBR which includes the previous vegetation density before the fire and minimizes the error caused by pre-fire vegetation. The RdNBR is defined as the relationship between the vegetation killed through the fire and the amount of pre-fire vegetation. Thus, positive RdNBR values represent a decrease in vegetation cover and consequently a high burn severity; negative values indicate an increase of vegetation and thus characterize unburned areas [36]. According to existing thresholds derived from measured data after fire events from Miller and Thode [13] four burn severity classes are defined from the RdNBR calculation.

Impacts of Forest Fires on Runoff

Background

Fire severity is the greatest determinant on the impact of a fire on streamflow generation. The changes in annual water budgets can last for decades in forested regions [39]. Severe fire events can reduce infiltration rates as a consequence of the generation of a hydrophobic soil layer caused by high temperatures or soil sealing forming crusts after a fire [10,39]. The decreased infiltration has an immense effect on post-fire runoff especially for extreme precipitation events during convective storms, when overland flow rises. The soils generally recover within a few years [39]. The long-term effects of forest fires on hydrology are caused by a reduction in evapotranspiration due to the loss of biomass, similar to those effects occurring after harvesting. Interception of rain and snow is also reduced. During the winter the snow melt rates are higher due to reduced shading by canopy [39].

The Southern Rockies Watershed Project [11] analyzed the impact of disturbance by the 2003 Lost Creek wildfire on hydrology, water quality and ecology on several reaches affected by the fire. The results are an increase in snow packs (54% higher in burned forest stands compared to unburned) and snow density, higher runoff values, especially in low flow periods during summer due to less evaporation and interception (by 14% compared to mean values from the 1990s), and peak runoffs are produced more rapidly in burned watersheds. Furthermore, an earlier spring streamflow peak was identified as well as a second peak during the highest precipitation in June. To analyze whether wildfire impacts at smaller reaches propagate in the runoff behavior of the whole Castle River watershed further investigation is carried out.

Data and Methods

To analyze whether the established WaSiM model can reproduce the observed runoff under changed land cover conditions after the wildfire, further analyses are conducted. The pixels identified as *burned* using the RdNBR method are now reclassified to depict the land cover conditions after a wildfire. Previously, land cover in the burned area was mainly coniferous forest which contributes immensely to evapotranspiration. Due to the lack of on-site information about the conditions after the fire, four replacement options are tested in the burned area: (1) no change; (2) barren soil; (3) grass and (4) shrubs, for which individual land use parameter settings are provided in the hydrological model. Specific values for albedo, stomata resistance, leaf area index and root depth must be provided to calculate evapotranspiration [20]. The parameter values for the four classes are presented in Table 1. According to Schulla [20] these are the most important parameters and the values vary between the four classes. It is notable, that barren land has mostly constant values as vegetation is missing.

First a model run from August 1993 to July 2003 is carried out with the original land cover to set the initial state of the hydrological conditions for the time the fire began. With these initial conditions, model runs for the first year after the fire from August 2003 to October 2004 are executed with the four land cover properties applied for the burned area. Subsequently, the different applied land cover classes in the burned areas are assessed regarding the model efficiency of the produced runoff.

Table 1. Selected default parameter values defined in the WaSiM landuse table for the four different land cover classes in the burned area. A seasonal evolution of vegetation can be defined with a high temporal resolution for the parameters in the table.

Parameter in (Landuse_Table)	Explanation	Previous Land Cover (Coniferous Forest)	Barren Land	Grass	Shrub
Albedo	Albedo	0.12	0.25	0.25	0.2
rsc (s/m)	Leaf surface resistance	220–320	400	200–360	200–320
rs_interception (s/m)	Evaporation of intercepted water	5	0.5	5	5
rs_evaporation (s/m)	Evaporation of water from the soil surface	1000	200	600	1000
LAI (1/1)	Leaf Area Index	6–10	1	2–3	3–5
Z0 (m)	Roughness length	3	1	0.03–0.04	0.2
VCF	Vegetation covered fraction	0.9–0.95	0.2	0.8–0.9	0.9–0.95
RootDepth (m)	Root depth	1.2	0.2	0.4	0.5

RESULTS AND DISCUSSION

Observed Climate and Streamflow Trend Results

AAFC data of monthly mean air temperatures (1950–2010) in the study basin range from −8.53 °C in January to 14.24 °C in July. The annual mean temperature is 2.66 °C with a standard deviation of 9.65 °C. Within the watershed there are extreme variations from these annual mean values as the elevation difference exceeds 1600 m, with measured values ranging from 4.51 °C (1213 m) to 0.01 °C (2139 m).

An indication of a warming climate is shown in Figure 2 by a linear increasing trend in yearly temperature by *ca.* 0.03 °C *i.e.*, 1 °C in 30 years. A general temperature increase is evident throughout the whole watershed. Temperatures in the lower section of Castle River watershed rose more considerably (2 °C) as compared to the mountainous parts (1 °C). Minimum and maximum temperatures are rising at a greater rate than the annual means which may indicate a shift towards less cold extremes and increasing hot day events.

Considering the monthly mean temperatures during the last 60 years the winter temperatures, especially in January have increased by about 4 °C in the last three decades. Monthly mean temperature data are shown in Figure 3. The spring temperatures also rose by *ca.* 2 °C in the 1990s and 2000s compared to the middle of the 20th century. In July, the monthly mean temperature in the 2000s was about 2 °C warmer than in the other decades and the August temperatures have also risen during these last decades. There is no distinct temperature trend visible in the fall.

The illustrated temporal and spatial trends of the temperature data from AAFC [18] do generally agree with the conclusions from the IPCC [40]. In the southern Rockies, especially temperatures between December and April have increased by up to 2.5 °C during the last six decades. This trend is also reflected by data from the Castle area in Figure 2.

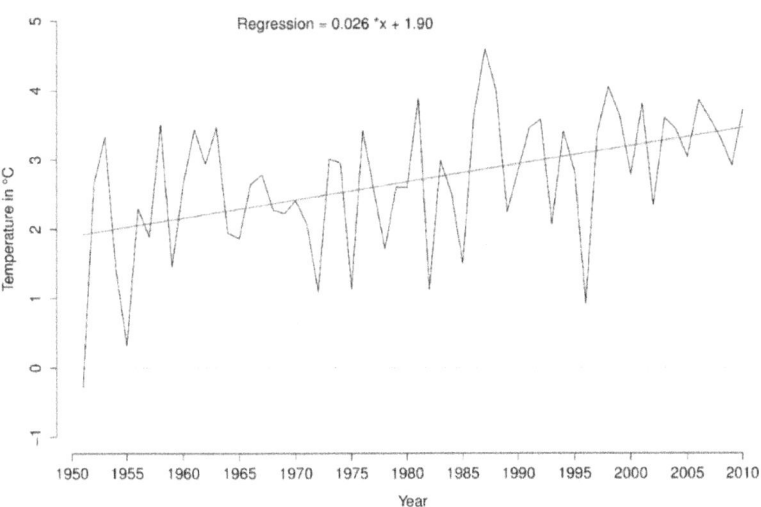

Figure 2. Annual mean temperatures between 1950 and 2010. The linear regression shows an increasing trend in temperatures.

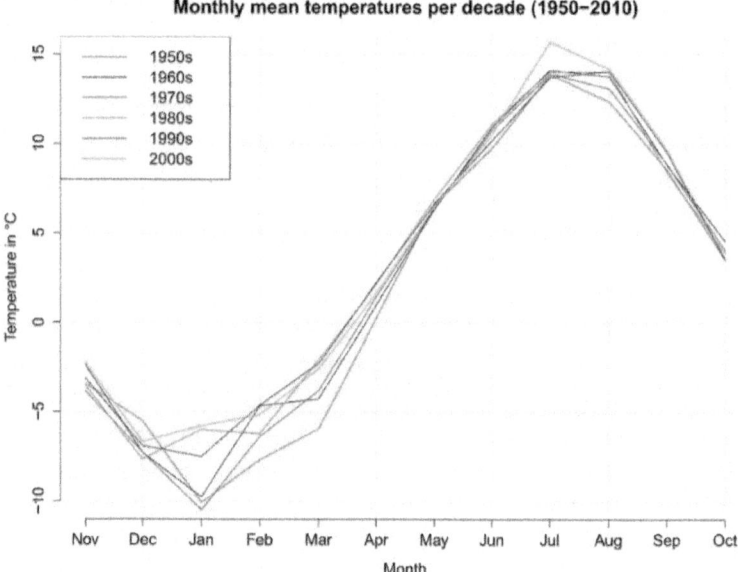

Figure 3. Monthly mean temperatures per decade (1950–2010).

Mean annual precipitation in the Castle River watershed is estimated at 686 mm·a^{-1} [18], with a high variability within the watershed ranging from *ca.* 600 mm·a^{-1} in the lower valleys to over 1000 mm·a^{-1} in the mountains [6]. Highest precipitation is observed in June (average of 86.8 mm) with a minimum in July (46.4 mm) or during winter [18].

Analyzing precipitation data from 1950 to 2010 reveals the following trends: The annual precipitation sum is generally slightly increasing with a high interannual variability. The seasonal distribution of precipitation is changing, even though the patterns are more ambiguous than the temperature data. Winter precipitation has been decreasing in the last six decades while the summer precipitation is rising with increasing temperatures. The 1980s showed significantly lower precipitation compared to other decades (annual mean: 655 mm·a^{-1} in the 1980s *vs.* 692 mm·a^{-1}). In the 1980s, winters, springs and summers were drier and warmer. The reasons for this relative drought were probably weaker west winds and a series of stable high pressure fields which blocked out storms [41].

Spatially distributed precipitation is high in the mountainous areas of the watershed in the south and west and is declining towards the northeast. The trend from 1950 to 2010 amplifies this pattern, as the mountains became more humid with increases of the annual precipitation sum by about 50 mm and the lower lying areas became drier.

The derived precipitation trend results are similar to those presented in the IPCC [40], which show an annual precipitation increase by 12% on average but in the southern Rockies this trend is weaker [42]. Significant changes can be observed in terms of temporal distribution of precipitation which conform to those of Castle watershed. Heavy precipitation frequencies increased as well in the last 60 years [42,43].

Considering gauge data at Beaver Mines from Environment Canada [15], the mean runoff has not been constant throughout the last 45 years. In the 1960s the mean annual flow was 15.1 $m^3 \cdot s^{-1}$ (580 $mm \cdot a^{-1}$), then it declined until, in the 1980s, it was 13.2 $m^3 \cdot s^{-1}$ (507 $mm \cdot a^{-1}$). In the 1990s it increased to 14.8 $m^3 \cdot s^{-1}$ (569 $mm \cdot a^{-1}$) and then decreased again to 13.3 $m^3 \cdot s^{-1}$ (511 $mm \cdot a^{-1}$) between 2000 and 2010. The monthly low and high flows show a corresponding behavior through the decades.

Overall, when applying a linear regression, the data shows a trend of decreasing annual mean runoff values over the considered decades. The observed annual mean runoff and the derived trend in a 5-year moving average for the whole period is shown in Figure 4. From 1962 to 1999, runoff decreased about 20%, which corresponds to an annual decrease of $ca.0.53\%$. This reduction of annual river flow conforms to the investigations from Byrne [1] who analyzed streamflow of the Castle and Oldman Rivers. They derived a general decline in mean streamflow of the Oldman River watershed by about 26% since 1949. The reason for this trend is attributed to the reduction of snow packs in the watershed due to the identified rising temperatures. Although precipitation in winter is increasing, this cannot compensate for the warming and thus more precipitation falls as rain than as snow. Due to less melting spring runoff is reduced [1], even at a stronger rate than in other comparable watersheds in the North American Rocky Mountain region.

Considering the runoff trend for the decade 2000 to 2010 a slight increase is detected. This can mainly be explained by extraordinary high precipitation in 2002 with 800 $mm \cdot a^{-1}$ and over 1050 $mm \cdot a^{-1}$ in the year 2005 [18].

Figure 5 shows the monthly mean runoff values per decade. The declining runoff values in May and June are an indicator of reduced snowpack, as less snow supplies less melt water for runoff generation. It is noticeable that the 1980s reflected a much lower runoff rate than the other decades, especially between May and August due to the relative dryness in this decade. Apart from the 1980s, the peak streamflow in June has constantly declined. In the other seasons, a distinct trend is not visible.

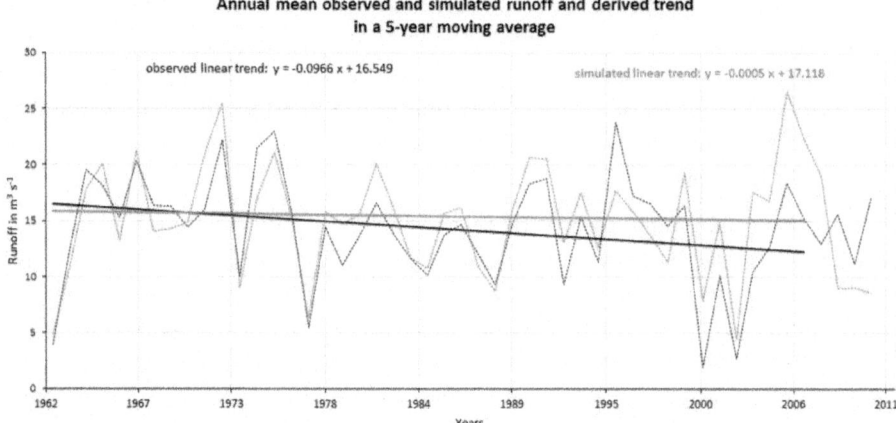

Figure 4. Annual mean values of observed and simulated streamflow at the gauge Beaver Mines at Castle River. The derived linear trends are calculated with a 5-year moving average from 1962 to 2010.

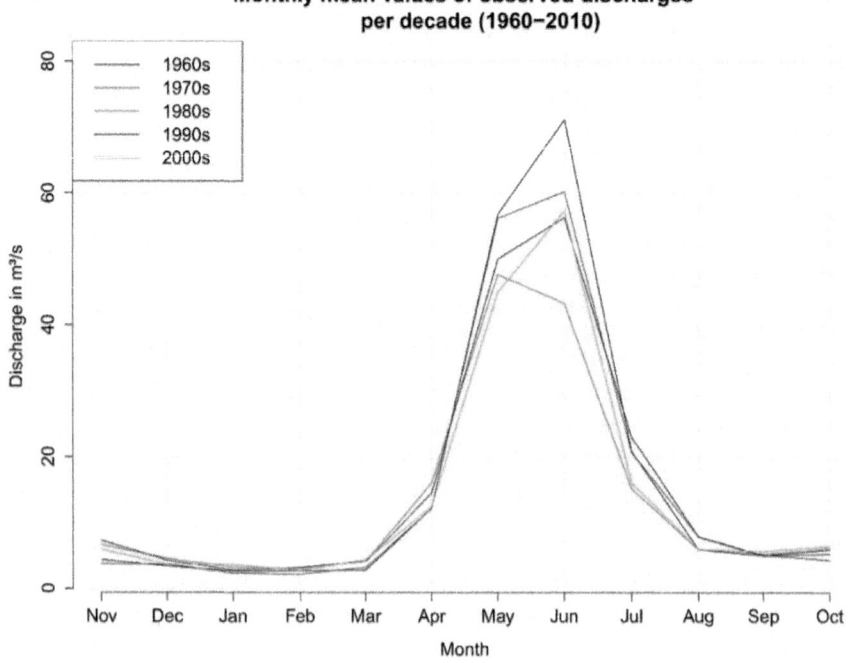

Figure 5. Monthly mean values of observed discharges per decade (1960–2010).

Calibration and Validation Results of WaSiM

First, time series of the simulated runoff in calibration and validation periods are discussed. Second, an analysis of the spatial output grids for water balance calculation and runoff components is executed.

Calibration and validation results are listed in Table 2. The result of the best calibration run from 1960 to 1970 is shown inFigure 6. Daily and monthly mean observed and simulated data are represented as well as the precipitation.

Model efficiency is assessed according to Moriasi [30] who suggested a performance rating in four NSE classes: (i) Very good ($0.75 < NSE \leq 1.00$); (ii) Good ($0.65 < NSE \leq 0.75$); (iii) Satisfactory ($0.50 < NSE \leq 0.65$) and (iv) Unsatisfactory ($NSE \leq 0.50$). Generally, the efficiency of the established WaSiM model is very good in daily (NSE: 0.81, R^2: 0.82) and in monthly mean temporal resolution (NSE: 0.88, R^2: 0.89), as is listed in Table 2. The mean annual flow is underestimated while mean high and low flows are overestimated. For the model validation run from 1970 to 1980, the accuracy is the same as in the calibration. R^2 also stays high between 1970 and 1999; from 1980 to 1990 it goes up to 0.87. It is conspicuous that the model efficiency decreases considerably to an NSE of 0.45. This result shows that a further investigation of modeling in this decade with the severe wildfire in 2003 has to be accomplished (3.3 and 4.5). The mean low flows are constantly overestimated in the validation periods. The mean high flows are overestimated but are modeled in the magnitude of the observed values.

Table 2. Calibration and Validation goodness of fit with MQ (mean runoff), MHQ (mean high flow) and MLQ (mean low flow) in $m^3 \cdot s^{-1}$ (first two years of initial time in model run are not considered in efficiency estimation).

	Period	Data	MQ	MQ Relative Error	MHQ	MLQ	NSE	R^2
Calibration:	1960–1970	observed	15.47		87	1.20		
		simulated	14.75	−5%	106	3.26	0.81	0.82
Validation:	1970–1980	observed	13.38		94	1.61		
		simulated	13.58	+1%	95	3.18	0.81	0.82
Validation:	1980–1990	observed	12.16		93	2.09		
		simulated	12.98	+6%	110	3.29	0.76	0.87
Validation:	1990–1999	observed	14.87		91	1.65		
		simulated	14.13	−5%	123	4.01	0.76	0.76
Validation:	2000–2010	observed	13.27		90	1.36		
		simulated	15.55	+15%	89	4.92	0.45	0.63

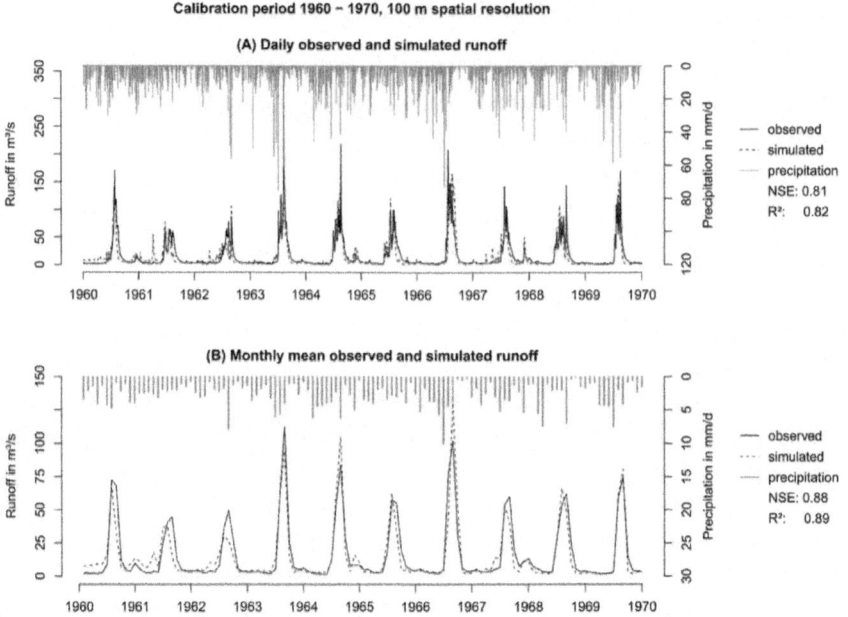

Figure 6. (**A**) Daily and (**B**) monthly observed and simulated runoff in the calibration period 1960–1970.

Generally the set up WaSiM model is able to reproduce runoff time series of the Castle River watershed and provides consistent simulation quality in a *ca.* 40 year period with very high accuracy.

In addition to the temporal and the quantitative accuracy of the model, spatial output data of the three runoff components is analyzed. Especially the direct flow but also the interflow is mainly generated in the mountainous parts, where precipitation is higher. Direct flow is very high in non-vegetated rocky summit regions. Base flow is mainly built in regions where the soil is more developed and is especially visible in the valleys. These spatial patterns seem to reflect real conditions very well. The interflow contributes the largest amount of water to the streamflow with over 440 mm·a⁻¹, whereas the direct flow and the base flow together only account for around 150 mm·a⁻¹. Base flow contributes the highest amount of runoff in the low flow periods. Interflow is especially high during snow melt in the spring to early summer whereas the direct flow reacts to high precipitation events.

A water balance [44] is calculated for the calibration period from 1960 to 1970 and presented in Figure 7. The mountainous precipitation is increased after the precipitation correction method in the model. According to Silins [11] who conducted a few meteorological measurements in the northern reaches of the

Castle watershed and detected annual precipitation values of over 1200 mm·a^{-1} the derived precipitation in the mountains can be claimed to be appropriate. However, the precipitation in the lower regions of the watershed is thereby overestimated. The mean annual precipitation at the meteorological station Beaver Mines, close to the gauge, amounts to around 650 mm·a^{-1} [24] whereas the model output defines at least 910 mm·a^{-1}. Nevertheless, the precipitation correction is applied, as without it the model efficiency is significantly lower (NSE: 0.40, R^2: 0.68) and the mean flow of the Castle River is significantly underestimated (5.5 m^3·s^{-1}compared to the observed 15.5 m^3·s^{-1}). Most of the runoff is produced in mountainous areas of the watershed, where precipitation correction delivers more appropriate values. Furthermore, most of the snow is produced in the mountains and the evapotranspiration has its maximum in the lower areas. Thus, the derived mistake by an overestimation of precipitation downstream can be neglected, as its contribution to the streamflow generation is low.

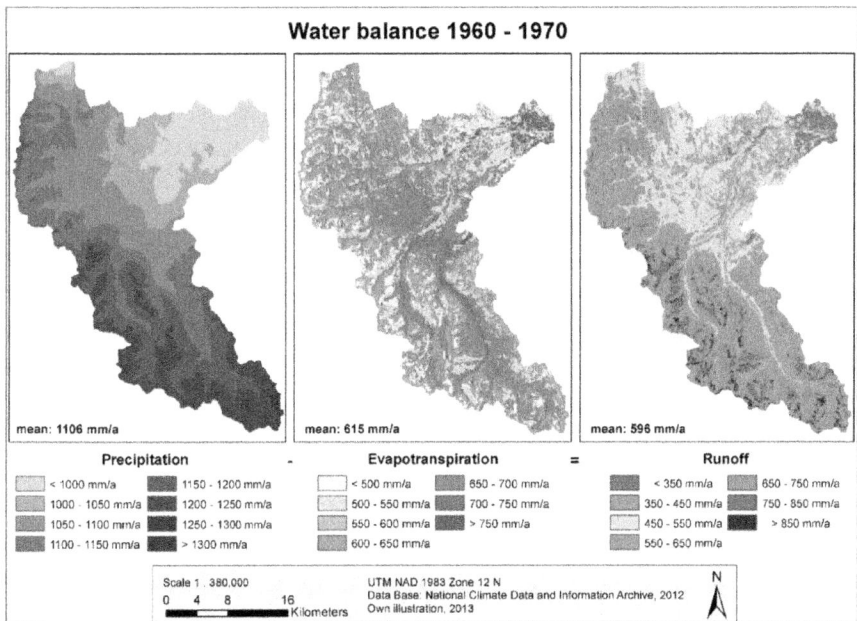

Figure 7. Simulated mean annual water balance in the calibration period from 1960 to 1970. The values are derived from the spatial output data of WaSiM and represent interpolated precipitation, as well as the modelled evaporation and runoff.

The evapotranspiration varies between values of *ca.* 500 mm·a^{-1} to more than 750 mm·a^{-1}. Due to low transpiration the calculated evapotranspiration is lower in non-vegetated areas such as the higher altitudes. In the forested valleys and in the agricultural used areas in the northeast, very high evapotranspiration values can be detected. The mean runoff for the period 1960 to 1970 is 596 mm·a^{-1}. It is obvious that the highest runoff values (up to over 850 mm·a^{-1}) are generated in the mountain tops as precipitation and thus snow cover is higher there. The lowest contribution to the runoff is generated in the agricultural areas because of lower precipitation and higher evaporation.

The results of the established WaSiM model show that it is able to reproduce runoff with high accuracy over a period of 40 years. This also implies that the partially calculated or generated input data is relatively accurate and that calibration has been successful in this study.

Comparison of Observed and Simulated Declining Runoff Trend

The observed and simulated runoff in an annual mean is demonstrated in Figure 4. Analyzing the trends in total runoff between 1960 and 1999, the observed data show a linear decrease of *ca.* 1 m^3·s^{-1} (38 mm) in 12 years. Therefore, in the 40 years considered, the observed streamflow declined by about 3.1 m^3·s^{-1} which equals a decrease of 119 mm·a^{-1}, although in the four decades the trend has not been continuous. Regarding monthly mean values the simulated runoff trend has a very similar inclination compared the observed. A decrease of 1 m^3·s^{-1} in 9 years and a total decline of 4.1 m^3·s^{-1} (157 mm·a^{-1}) between 1960 and 1999 can be derived from the simulation. Low and high flow trends are also decreasing. The decline in snow cover since 1960 is captured by the model. WaSiM generates a total snow storage output which sums the snow in mm per day. The snow cover is varying but the linear trend shows a distinct decline of 53 mm·a^{-1} in an annual mean.

Considering the observed and simulated runoff between 2000 and 2010 a different behavior of the model is evident. Runoff is slightly increasing in observed values while WaSiM produces a decrease of 9.1 m^3·s^{-1} in the decade when regarding monthly mean values. Figure 4 shows that between 2002 and 2007 runoff is overestimated considerably. The assumption is that the changed condition in the basin after the 2003 wildfire is the reason for the decline in model efficiency (compare Table 2). Thus, further investigation in this decade is conducted in chapter 4.5. The water balance for the period 1990 to 1999 is calculated and then compared to the water balance from 1960 to 1970. The change maps for simulated precipitation, evapotranspiration and runoff are demonstrated in Figure 8. Considering precipitation, a spatially differenced distribution and an increased amplitude is evident, as the decrease in the lower parts is up to 30 mm·a^{-1}, and the increase, especially in the northern and western mountains, is up to 99 mm·a^{-1}. This observation conforms with the projections from Field [42] which show an increasing variability in precipitation.

Considering evapotranspiration values between the 1960s and 1990s, an overall increase can be seen due to increased temperatures. Nevertheless, in the higher mountainous parts a decreasing evapotranspiration was simulated. A possible explanation for this development could be the decline in snow cover which is represented by the model. Mac Donald, *et al.* [45] emphasized the importance of sublimation of snow cover for the mass balance in a watershed. Therefore, the reduction in evapotranspiration could possibly be explained by the loss of sublimation which contributes to evapotranspiration. As a consequence of the developments in the distribution of precipitation and evapotranspiration, runoff is on average declining between the 1960s and 1990s. In the western, high altitude regions, more precipitation produces more runoff, especially direct streamflow and interflow. In the valleys, increasing evapotranspiration reduces the runoff generation and causes its declining trend.

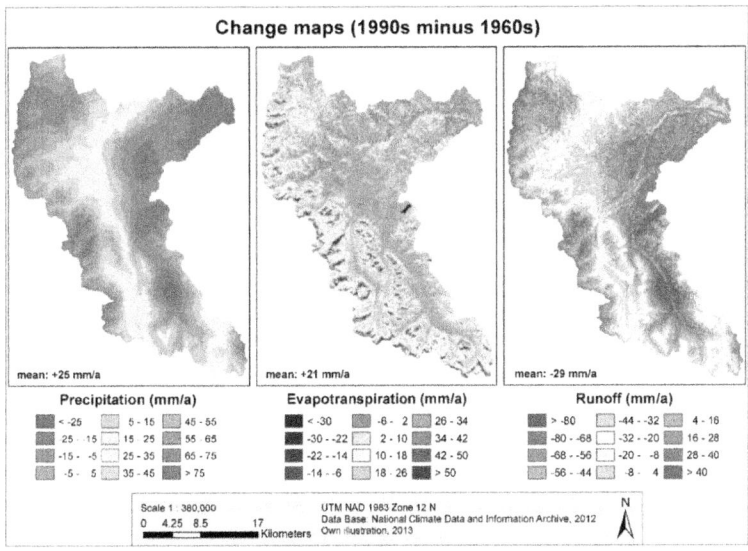

Figure 8. Change maps of precipitation, evapotranspiration and runoff generation between the 1960s and the 1990s.

It can be claimed that WaSiM is able to reproduce quite well runoff values under changing climate conditions. The decline in snow cover and, consequently, in total runoff is reproduced in the model results. Changes in precipitation distribution and quantity, an increase in evapotranspiration and trends in runoff generation in which mainly the interflow is declining are demonstrated with the established model. The intensity of decline is slightly overestimated. Further investigation with measured input data or a tested and approved precipitation correction factor would be necessary to describe the changes to the watershed more accurately.

Results of Burn Ratio and Derived Burned Areas

The result of the Relative differenced NBR (RdNBR) is presented in Figure 9. High RdNBR values signify a decreased vegetation cover and thus a high burn severity. Low or negative RdNBR values indicate an increase in vegetation cover and show unchanged areas [13]. The derived image shows a clear zoning of the Lost Creek fire mainly in the high severity class.

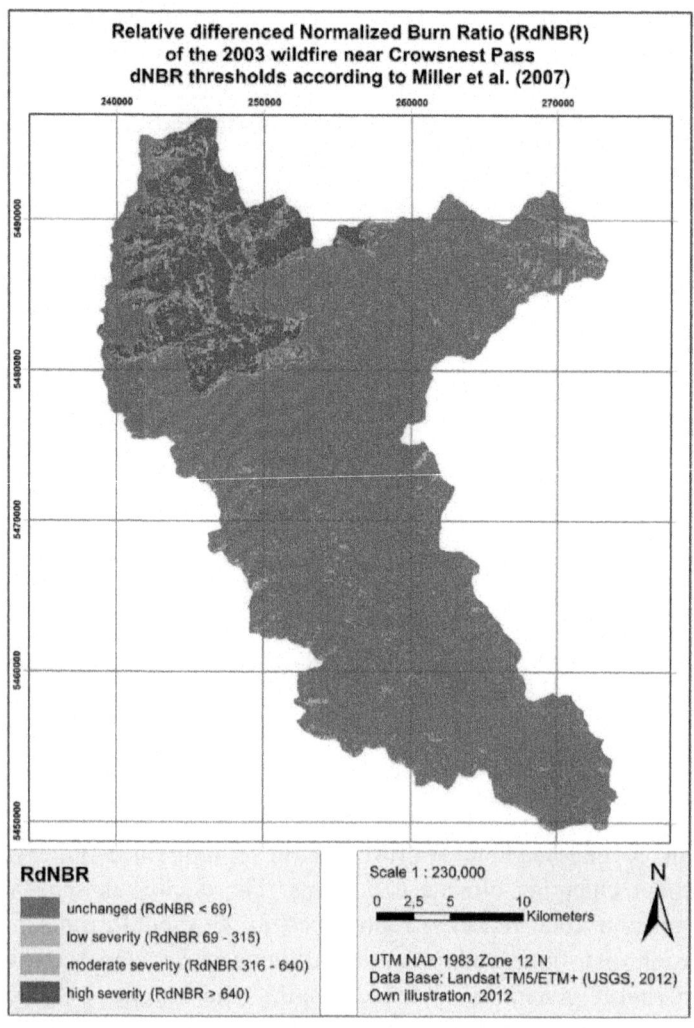

Figure 9. Relative differenced Normalized Burn Ratio (RdNBR) of the Lost Creek fire; thresholds according to [13].

Field measurements and derived burn severity values for the wildfire in the Castle River watershed are not available. Hence, the dNBR and RdNBR thresholds derived from Miller and Thode [13], who analyzed 14 fires in the Sierra Nevada Mountain range in California, USA, are applied. They compared measured CBI (Composite Burn Index) values with dNBR and RdNBR from pre- and post-fire Landsat data and defined thresholds, derived from a regression model.

Applying the RdNBR shows that the derivation of burned areas from Landsat satellite imagery is generally feasible in the Castle River watershed. The area derived by this method agrees quite well with the size collected by ESRD Alberta [33], with a discrepancy of 14%. There are two possible reasons for this difference. Either, the burned area was collected very generally and the RdNBR calculation gives a more precise result for the burned areas, or else there are mistakes within the calculation or the application of the thresholds from [13] are used. Therefore, the definition of burned areas in the severity categories can only be an approximation. For a correct delineation of burn severity categories in Castle River watershed, location-dependent CBI-values would have been necessary. Nevertheless, the presented method has several advantages. The Landsat data are freely available, and long time-series exist so that change detection and post-fire succession analyses are possible. Compared to the simple burned area map from ESRD Alberta [33], Landsat data deliver valuable information about the plant characteristics and changes of vital vegetation occurring with a fire event [46]. Most importantly, the RdNBR method delivers information about different grades of burn severity which are, according to Luce [39], the supreme determinants for the dimension of the impact of a wildfire on the hydrology of a watershed. For further analysis of the impacts of the extensive forest fire on the hydrology of Castle River watershed, moderate and high severity RdNBR values are included in the land cover map and defined in a "burned" land cover class. These newly derived land cover maps are later applied as input data sets for the hydrological model WaSiM.

Results of Simulated Runoff Reaction to Fires

The model run from August 1993 to July 2003, with the purpose to produce an initial state for the time the fire began, does not have a considerable lower efficiency than the model runs in the decades before (NSE: 0.76, R^2: 0.82). Simulating runoff after the fire from August 2003 to November 2004 does not lead to accurate results applying the original land cover input for the burned area (NSE: 0.51, R^2: 0.86). The mean annual flow is overestimated considerably by 18%. Thus, it can be assumed that the model allows an accurate runoff simulation only until the fire changed the land cover conditions.

Applying the three other land cover characteristics in the burned area as explained in 3.4.2, NSE efficiency results range between 0.44 with shrub and 0.56 for barren soil characteristics in the burned area. The relative error of the mean annual runoff is around 20%. Barren soil characteristics in the burned area deliver the best results for this time period (NSE: 0.56, R^2: 0.90, MQ relative error: +17%). The simulation results compared with the observed data are shown in Figure 10. Although the mean annual runoff is overestimated under this scenario, satisfactory results can still be achieved, regarding the categorization from Moriasi [30]. Consequently, barren soil characteristics seem to represent a burned vegetation in the first year after the fire more accurately than the other soil characteristics applied. Although dead tree trunks and burned organic litter are present in burned areas, all vegetation based hydrological processes are obviously interrupted. Especially low flow is represented quite well. Melt runoff peaks are characterized quite well but the runoff in July, when highest precipitation values are present, is overestimated by WaSiM.

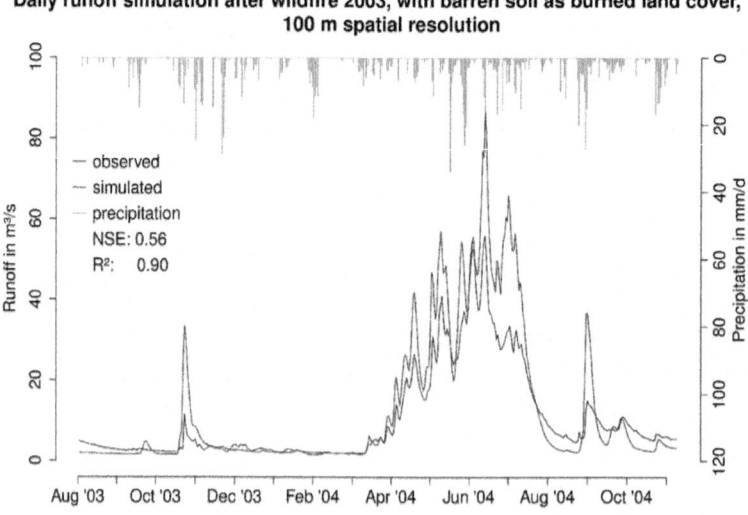

Figure 10. Daily runoff simulation after the wildfire in summer 2003. This graph represents the model run from August 2003 to November 2004 with barren soil as burned land cover classification.

Regarding the simulated total runoff, it is evident that in the burned region runoff generation is slightly increased compared to the unburned area. This is caused by an increase of interflow by *ca.* 8% in the burned area. Due to an absent transpiration without vegetation cover, more water can infiltrate and contribute to the total runoff. This is especially caused by the parameterization of low LAI, low evaporation (rs_interception and rs_evaporation) and high rsc values which are demonstrated in Table 1.

Despite the modifications it is conspicuous that WaSiM still produces less accurate results for the period after the forest fire than in the four decades before. A reason for this can be traced to the change in land cover after the fire. In addition, as the model results showed a reaction to changes in the burned area land cover parameterization it demonstrates that this approach is correct. Hence a new land cover class for burned areas needs to be established.

Another reason why a high model accuracy cannot be achieved for the burned simulation scenario is the climate variability. As shown before, WaSiM reproduces a declining runoff trend between 1960 and 2010 due to changes in climate patterns, but the runoff and the decreasing trend are also overestimated in this period. In the 2000s, precipitation was lower in the winter months but higher in the summer months than in other decades. The applied precipitation correction likely resulted in an overestimation of precipitation in summers in the watershed, which, in turn, alters the annual streamflow and unrealistically increases the peak flows in June and July.

Therefore, generally, a superposed situation is given in Castle River watershed after the 2003 fire. On the one hand a distinct climate trend reduces the freshets, which mainly contribute to runoff generation in southern Albertan alpine catchments. On the other hand, the general behavior after wildfires is an increase in runoff which results in an earlier spring melt flow as well as higher peak flows. Both characteristics can be detected in the observed and modeled runoff values. It is demonstrated by a higher low flow between August and March and an earlier begin of melt possibly due to the fire effects as well as a significantly decreased peak flow as a consequence of reduced snow packs due to climate variability.

CONCLUSIONS AND OUTLOOK

This study investigated the hydrology of Castle River watershed in the southern Albertan Rocky Mountains with particular focus on climate change induced alterations and the impacts of forest fires on streamflow. The initially raised questions and objectives could mainly be achieved. The declining runoff trend, predominantly a consequence of higher evapotranspiration and decreasing snow cover due to rising temperatures, is evident in the Castle River. This trend could also be reproduced by the applied hydrological model WaSiM, which resulted in high accuracy in runoff simulations in the watershed when compared to recorded streamflow records.

The fire severity of the Lost Creek fire in 2003 was detected with pre- and post-fire Landsat satellite images using a differencing burn ratio method (RdNBR). Thus, the area burned by the fire as well as severity classes could be defined very precisely. The impact on the discharge in the watershed is evident in the observed runoff. Nevertheless, a decrease in spring peak flow due to climate induced decreasing snow cover was dominant over the increased peak flow as a consequence of the fire.

Applying the model after the wildfire in 2003 lead to a significantly lower model efficiency. This concludes that WaSiM output is sensitive to land cover changes due to forest fires. An attempt to imitate burned land cover characteristics resulted in increasing accuracy in runoff simulations. Though it is anticipated that a new land cover class which contains parameter values more representative of a burned forest would result in an improved simulation of runoff with WaSiM after a fire. Depending on burn severity a land cover class similar to barren soil would be appropriate, but including features such as e.g. a burned organic layer, changed soil characteristics and dead tree trunks which can still intercept some water. In the following years a dynamic forest succession in the burned areas has to be applied.

This study also shows that it is possible to undertake satisfactory hydrological modeling even if principle input data is missing. The meteorological input data relative humidity and global radiation were estimated based on air temperature and precipitation data, soil information in the watershed was deviated from land use and the temperature and precipitation data were also not measured but interpolated in a broad scale. This confirms the quality of the model WaSiM, which yields plausible results with a coherent parameterization even with poor input data availability.

Further investigation of the outlined research questions is recommended. Future trends are tending towards lower discharge due to decreasing snow cover as a result of climate change [1] as the trend from 1960 to 2010 already represents. Furthermore, an increase in forest fires is expected due to rising temperatures and an extended fire season length [47]. Additionally, changes in land cover either by forest fires or other partly anthropogenic interventions will impact the hydrology in the Castle area in the future. The establishment of gauges of sub-basins and meteorological stations at different altitudes in the watershed would provide significant hydro-meteorological data that is currently estimated [18]. A site-specific elevation dependent regression for correcting measured precipitation could be derived and applied in the hydrological model to obtain spatially distributed meteorological information. A precise soil texture mapping would possibly further improve the conditions to work with the model and produce more accurate results as infiltration processes could be reproduced more precisely.

However, the presented methodological approach is considered robust to implement a scenario simulations framework for projecting the impacts of future climate and land cover change in the vulnerable region of Alberta's Rocky Mountains and comparable mountainous watersheds.

ACKNOWLEDGMENTS

Our thanks go to the Department of Geography, Ludwig-Maximilians-Universitaet Muenchen, and to the Department of Geography, University of Lethbridge, for providing data, methodical support and a working environment with the required facilities and programs. We would also like to mention the enriching cooperation between the two universities and in the project ABBY-Net. The idea for this study was initiated during an ABBY-Net summer school in Alberta.

AUTHOR CONTRIBUTIONS

Johanna Springer carried out all methodical works as well as analyses, assessment and discussion of the results and wrote the initial manuscript. Ralf Ludwig and Stefan Kienzle provided guidance with their expertise in hydrology, climate data analysis, remote sensing and hydrological modeling. Furthermore, they contributed by reviewing and completing the text.

REFERENCES

1. Byrne, J.; Kienzle, S.; Johnson, D.; Duke, G.; Gannon, V.; Selinger, B.; Thomas, J. Current and future water issues in the Oldman River Basin of Alberta, Canada. *Water Sci. Technol.* **2006**, *53*, 327–334.

2. Forbes, K.A.; Kienzle, S.W.; Coburn, C.A.; Byrne, J.M.; Rasmussen, J. Simulating the hydrological response to predicted climate change on a watershed in southern Alberta, Canada. *Clim. Chang.* **2011**, *105*, 555–576.

3. Lapp, S.; Byrne, J.; Townshend, I.; Kienzle, S. Climate warming impacts on snowpack accumulation in an alpine watershed. *Int. J. Climatol.* **2005**, *25*, 521–536.

4. Kienzle, S.W. Water Yield and Streamflow Trend Analysis for Alberta Watersheds. Available online: http://www.albertawater.com/index. php/projects-research/dynamics-of-alberta-s-water-supply/41-water-research/dynamics-of-alberta-s-water-supply/518-water-yield-and-

streamflow-trend-analysis-for-alberta-s-watersheds (accessed on 1 March 2013).

5. Rood, S.B.; Samuelson, G.M.; Weber, J.K.; Wywrot, K.A. Twentieth-century decline in streamflows from the hydrographic apex of North America. *J. Hydrol.* **2005**, *306*, 215–233.

6. Sheppard, D.H.; Parkstrom, G.; Taylor, C. Bringing It Back: A Restoration Framework for the Castle Wilderness. Available online: http://www.ccwc.ab.ca/files/BRINGING_IT_BACK.pdf (accessed on 17 December 2012).

7. Wulder, M.A.; White, J.C.; Alvarez, F.; Han, T.; Rogan, J.; Hawkes, B. Characterizing boreal forest wildfire with multi-temporal Landsat and LIDAR data. *Remote Sens. Environ.* **2009**, *113*, 1540–1555.

8. Stocks, B.J.; Mason, J.A.; Todd, J.B.; Bosch, E.M.; Wotton, B.M.; Amiro, B.D.; Flannigan, M.D.; Hirsch, K.G.; Logan, K.A.; Martell, D.L.; *et al.* Large forest fires in Canada, 1959–1997. *J. Geophys. Res.* **2003**.

9. Flannigan, M.D.; Logan, K.A.; Amiro, B.D.; Skinner, W.R.; Stocks, B.J. Future area burned in Canada. *Clim. Chang.***2005**, *72*, 1–16.

10. Pierson, F.B.; Robichaud, P.R.; Spaeth, K.E. Spatial and temporal effects of wildfire on the hydrology of a steep rangeland watershed. *Hydrol. Process.* **2001**, *15*, 2905–2916.

11. Silins, U.B.; Kevin, D.; Stone, M.; Emelko, M.B.; Boon, S.; Williams, C.; Wagner, M.J. Howery, Jocelyn Southern Rockies Watershed Project: Impact of Natural Disturbance by Wildfire on Hydrology, Water Quality, and Aquatic Ecology of Rocky Mountain Watersheds Phase I (2004–2008) Final Report. Available online: http://oldmanbasin.org./files/Publications/SouthernRockiesWatershedProject-FinalReport-(Phase-I-2004-2008).pdf (accessed on 17 December 2012).

12. Wainwright, J.; Mulligan, M. *Environmental Modelling: Finding Simplicity in Complexity*; Wiley: Chichester, UK, 2004.

13. Miller, J.D.; Thode, A.E. Quantifying burn severity in a heterogeneous landscape with a relative version of the delta Normalized Burn Ratio (dNBR). *Remote Sens. Environ.* **2007**, *109*, 66–80.

14. Silins, U.; Stone, M.; Emelko, M.B.; Bladon, K.D. Sediment production following severe wildfire and post-fire salvage logging in the Rocky Mountain headwaters of the Oldman River Basin, Alberta. *CATENA* **2009**, *79*, 189–197.

15. Environment Canada. Water Survey of Canada-Castle River near Beaver Mines: Station ID 05AA022. Available online:http://www.wsc.ec.gc.ca/

applications/H2O/report-eng.cfm?yearb=&yeare=&station=05AA022& report=daily&year=2011 (accessed on 17 December 2012).

16. Alberta Wilderness Association. Areas of Concern-Castle. Available online:http://albertawilderness.ca/issues/wildlands/areas-of-concern/ castle (accessed on 10 December 2012).

17. Kulig, J.C.; Edge, D.; Reimer, W.; Townshend, I.; Lightfoot, N. Levels of Risk: Perspectives of the Lost Creek Fire. Available online:https://www. uleth.ca/dspace/bitstream/handle/10133/1266/Levels%20of%20Risk. pdf?sequence=1(accessed on 17 December 2012).

18. Agriculture and Agri-Food Canada. Gridded Climate Data. Available online: http://www4.agr.gc.ca/AAFCAAC/displayafficher. do?id=1227620138144&lang=eng (accessed on 10 October 2012).

19. University of Waterloo. Climate Dataset (Daily 10km Grids). Available online: http://www.lib.uwaterloo.ca/locations/umd/digital/documents/ ClimateDatasetDaily10kmGrids.html (accessed on 25 February 2013).

20. Schulla, J. Model Description WaSiM (Water Balance Simulation Model). Available online: http://www.wasim.ch/downloads/doku/wasim/ wasim_2012_ed2_en.pdf (accessed on 17 December 2012).

21. Jasper, K.; Calanca, P.; Gyalistras, D.; Fuhrer, J. Differential impacts of climate change on the hydrology of two alpine river basins. *Clim. Res.* **2004**, *26*, 113–129.

22. Beckers, J.; Smerdon, B.; Wilson, M. Review of hydrologic models for forest management and climate change applications in British Columbia and Alberta. In *FORREX Forum for Research and Extension in Natural Resources, Kamloops*; Forrex Forum for Research and Extension in Natural Resources: Kamloops, BC, Canada, 2009; Forrex Series 25.

23. Klein Tank, A.M.G.; Wijngaard, J.B.; Können, G.P.; Böhm, R.; Demarée, G.; Gocheva, A.; Mileta, M.; Pashiardis, S.; Hejkrlik, L.; Kern-Hansen, C.; *et al.* Daily dataset of 20th-century surface air temperature and precipitation series for the European Climate Assessment. *Int. J. Climatol.* **2002**, *22*, 1441–1453.

24. National Climate Data Information Archive Climate Data Online. Available online: http://climate.weatheroffice.gc.ca/climateData/ canada_e.html (accessed on 1 January 2013).

25. Thornton, P.E.; Hasenauer, H.; White, M.A. Simultaneous estimation of daily solar radiation and humidity from observed temperature and precipitation: An application over complex terrain in Austria. *Agric. For. Meteorol.* **2000**,*104*, 255–271.

26. Bristow, K.L.; Campbell, G.S. On the relationship between incoming solar radiation and daily maximum and minimum temperature. *Agric. For. Meteorol.* **1984**, *31*, 159–166.

27. GeoBase Canadian Council on Geomatics. Digital Elevation Model Data. Available online: http://www.geobase.ca/geobase/en/find.do?produit=cded (accessed on 2 November 2012).

28. GeoBase National Land Cover Project Team; DB Geoservices Inc. GeoBase Land Cover Product: User Needs Assessment. Available online:http://www.geobase.ca/geobase/en/data/landcover/csc2000v/description.html (accessed on 5 February 2013).

29. CSSS Canadian Society of Soil Sciences; Department of Soil Science, University of Saskatchewan. Soils of Canada: Orders. Available online:http://www.soilsofcanada.ca/orders/index.php (accessed on 2 February 2013),.

30. Moriasi, D.N.; Arnold, J.G.; van Liew, M.W.; Bingner, R.L.; Harmel, R.D.; Veith, T.L. Model evaluation guidelines for systematic quantification of accuracy in watershed simulations. *Trans. ASABE* **2007**, *50*, 885–900.

31. Nash, J.E.; Sutcliffe, J.V. River flow forecasting through conceptual models, Part I-A discussion of principles. *J. Hydrol.* **1970**, *10*, 282–290.

32. Krause, P.; Boyle, D.P.; Bäse, F. Comparison of different efficiency criteria for hydrological model assessment. *Adv. Geosci.* **2005**, *2005*, 89–97.

33. ESRD Alberta Spatial Wildfire Data: Historical Wildfire Perimeter Data: 1931–2011. Available online: http://srd.alberta.ca/Wildfire/WildfireStatus/HistoricalWildfireInformation/SpatialWildfireData.aspx (accessed on 2 February 2013).

34. Soverel, N.O.; Perrakis, D.D.B.; Coops, N.C. Estimating burn severity from Landsat dNBR and RdNBR indices across western Canada. *Remote Sens. Environ.* **2010**, *114*, 1896–1909.

35. Key, C.H.; Benson, N.C. Landscape assessment: Remote sensing of severity, the normalized burn ratio. In *FIREMON: Fire Effects Monitoring and Inventory System: General Technical Report*; Lutes, D.C., Ed.; USDA Forest Service, Rocky Mountain Research Station: Ogden, UT, USA, 2006; Volume 2006, pp. 219–273.

36. Miller, J.D.; Knapp, E.E.; Key, C.H.; Skinner, C.N.; Isbell, C.J.; Creasy, R.M.; Sherlock, J.W. Calibration and validation of the relative differenced Normalized Burn Ratio (RdNBR) to three measures of fire severity in the Sierra Nevada and Klamath Mountains, California, USA. *Remote Sens. Environ.* **2009**, *113*, 645–656.

37. Epting, J.; Verbyla, D.; Sorbel, B. Evaluation of remotely sensed indices for assessing burn severity in interior Alaska using Landsat TM and ETM+. *Remote Sens. Environ.* **2005**, *96*, 328–339.

38. Verbyla, D.L.; Kasischke, E.S.; Hoy, E.E. Seasonal and topographic effects on estimating fire severity from Landsat TM/ETM+ data. *Int. J. Wildland Fire* **2008**, *17*, 527–534.

39. Luce, C.H. Land use and land cover effects on runoff processes: Fire. In *Encyclopedia of Hydrological Sciences*; Anderson, M.G., McDonnel, J.J., Eds.; Wiley: Hoboken, NJ, USA, 2005; Volume 3, pp. 1831–1837.

40. IPCC. *Climate Change 2007-Impacts, Adaptation and Vulnerability: Working Group II Contribution to the Fourth Assessment Report of the Intergovernmental Panel on Climate Change*; Parry, M.L., Canziani, O.F., Palutikof, J.P., van der Linden, P.J., Hanson, C.E., Eds.; IPCC Secretariat: Geneva, Switzerland, 2007.

41. Nkemdirim, L.; Weber, L. Comparison between the Droughts of the 1930s and the 1980s in the Southern Prairies of Canada. *J. Clim.* **1999**, *12*, 2434–2450.

42. Field, C.B.M.; Brklacich, M.; Forbes, D.L.; Kovacs, J.A.; Running Steven, W.; Scott, M.J. Chapter 14: North America. In*Climate Change 2007-Impacts, Adaptation and Vulnerability: Working Group II Contribution to the Fourth Assessment Report of the Intergovernmental Panel on Climate Change*; Parry, M.L., Canziani, O.F., Palutikof, J.P., van der Linden, P.J., Hanson, C.E., Eds.; IPCC Secretariat: Geneva, Switzerland, 2007.

43. Lemmen, D.S.; Warren, F.J.; Lacroix, J.; Bush, E. *From Impacts to Adaptation: Canada in a Changing Climate 2007*; Government of Canada: Ottawa, ON, Canada, 2007; Available online: http://www.nrcan.gc.ca/sites/www.nrcan.gc.ca.earthsciences/files/pdf/assess/2007/pdf/full-complet_e.pdf (accessed on 5 February 2013).

44. Brutsaert, W. *Hydrology: An Introduction*; Cambridge University Press: Cambridge, UK/New York, NY, USA, 2005.

45. MacDonald, M.K.; Pomeroy, J.W.; Pietroniro, A. On the importance of sublimation to an alpine snow mass balance in the Canadian Rocky Mountains. *Hydrol. Earth Syst. Sci.* **2010**, *14*, 1401–1415.

46. Roeder, A.; Hill, J.; Duguy, B.; Alloza, J.; Vallejo, R. Mapping fire events and post fire succession using long time-series of Landsat-TM and -MSS data. In Proceedings of the 5th International Workshop on Remote Sensing and GIS Applications to Forest Fire Management: Fire Effects Assessment, Zaragoza, Spain, 16–18 June 2005; La Riva Fernández,

J.R.D., Pérez Cabello, F., Chuvieco Salinero, E., Eds.; Universidad de Zaragoza, Servicio de Publicaciones: Zaragoza, Spain, 2005; pp. 287–290.

47. Gillett, N.P. Detecting the effect of climate change on Canadian forest fires. *Geophys. Res. Lett.* **2004**, *31*, L18211.

Chapter 12

THE USE OF H-SAF SOIL MOISTURE PRODUCTS FOR OPERATIONAL HYDROLOGY: FLOOD MODELLING OVER ITALY

Christian Massari[1], Luca Brocca[1], Luca Ciabatta[1], Tommaso Moramarco[1], Simone Gabellani[2], Clement Albergel[3], Patricia De Rosnay[3], Silvia Puca[4], and Wolfgang Wagner[5]

[1]Research Institute for Geo-Hydrological Protection, National Research Council CNR, Via di Madonna Alta 126, 06128 Perugia, Italy

[2]International Centre on Environmental Monitoring (CIMA) Research Foundation, Via A. Magliotto, 2-17100 Savona, Italy

[3]European Centre for Medium-Range Weather Forecasts (ECMWF), Shinfield Park, RG2 9AX Reading, UK

[4]Italian Civil Protection Department, Via Vitorchiano 2, 00189 Rome, Italy

[5]Department of Geodesy and Geoinformation, Vienna University of Technology, 1040 Vienna, Austria

ABSTRACT

The ever-increasing availability of new remote sensing and land surface model datasets opens new opportunities for hydrologists to improve flood forecasting systems. The current study investigates the performance of two operational soil moisture (SM) products provided by the "EUMETSATSatellite Application Facility in Support of Operational Hydrology and Water Management" (H-SAF, http://hsaf.meteoam.it/) within a recently-developed hydrological model called the "simplified continuous rainfall-runoff model" (SCRRM) and the possibility of using such a model at an operational level. The model uses SM datasets derived from external sources (*i.e.*, remote sensing and land surface models) as input for calculating the initial wetness conditions of the catchment prior to the flood event. Hydro-meteorological data from 35 Italian catchments ranging from 800 to 7400 km^2 were used for the analysis for a total of 593 flood events. The results show that H-SAF operational products used within SCRRM satisfactorily reproduce the selected flood events, providing a median Nash–Sutcliffe efficiency index equal to 0.64 (SM-OBS-1) and

0.60 (SM-DAS-2), respectively. Given the results obtained along with the parsimony, the simplicity and independence of the model from continuously-recorded rainfall and evapotranspiration data, the study suggests that: (i) SM-OBS-1 and SM-DAS-2 contain useful information for flood modelling, which can be exploited in flood forecasting; and (ii) SCRRM is expected to be beneficial as a component of real-time flood forecasting systems in regions characterized by low data availability, where a continuous modelling approach can be problematic.

INTRODUCTION

Flooding has become one of the events producing the most fatalities annually [1]. Whereas much progress has been made in meteo-forecasts and warnings and also in public preparedness, a comparable system for "quantitatively" predicting floods has experienced less progress, especially in floods occurring in medium–small catchment sizes (100–1000 km²), as demonstrated by recent events occurring in Italy (*i.e.*, Liguria, Tuscany and Sicily at the end of 2011, Umbria in 2012 and Sardinia in 2014).

Indeed, the anticipation of the magnitude of an event is crucial for performing the correct actions within civil protection activities, but this is not a simple task. First, the amount of precipitation that transforms an otherwise ordinary rainfall event into an extraordinary one is led by complex interactions between meteorology and hydrology, such as, among other important factors, the soil moisture (SM) conditions prior to the flood event [2,3,4,5]. Second, predicting a flood event is not only a matter of being able to correctly describe such factors, but it is strongly related to the capability of the early warning system in terms of the data and tools upon which it can rely. That is: (i) an appropriate rainfall-runoff (RR) hydrological model able to infer, with a certain degree of accuracy, the discharge hydrograph; and (ii) a dense network of sensors able to provide good quality observations in near real time.

These two requirements are not independent of each other. Indeed, the choice of the most appropriate RR model (e.g., continuous *versus* event-based models) often stems from the availability of certain types of data (e.g., evapotranspiration or temperature data) and from the number of sensors available in the catchment, which leads the hydrologist to the choice of a continuous, event-based, distributed or lumped hydrological model.

Generally, continuous models try to describe the different hydrological processes able to generate runoff, but require long-term and uninterrupted time series of rainfall and evapotranspiration data. This could be a strong limitation in poorly-gauged areas, mainly if hourly observations are needed [6]. On the other hand, event-based RR models are very appealing and frequently employed within operational flood forecasting systems [3], because of their simplicity, the need for reduced parametrization and data records (*i.e.*, only rainfall recorded during the event is needed) and last, but not least, the much lower computational demand. However, a major limitation of event-based models lies in the definition of the initial SM conditions of the catchment, which may strongly vary from one storm event to another [3,5,7], especially in regions characterized by strong seasonality, such as Mediterranean countries [8].

In the last few decades, many studies [4,5,9,10,11,12,13] have mentioned the high importance of SM, because of its ability to determine the partitioning of rainfall into runoff and infiltration [14]. In particular, the authors have demonstrated that when SM is characterized by high variability throughout the year (e.g., in Mediterranean regions), it represents, more than others quantities, a good proxy of the antecedent wetness conditions of the catchment, thus allowing better prediction of the runoff response to the rainfall input.

At the same time, the ever-increasing availability of SM measurements from *in situ* stations (International Soil Moisture Network (ISMN) [15]), satellite sensors (e.g., the Advanced Scatterometer (ASCAT), [16], the Advanced Microwave Scanning Radiometer for Earth observation (AMSR-E) and its successor, AMSR-2 [17], and the Soil Moisture and Ocean Salinity Mission (SMOS) [18]) and land surface models [19], at increasing temporal and spatial resolutions [20], has opened new possibilities for integrating such measurements into hydrological models, even at an operational level. By way of example, the "EUMETSAT Satellite Application Facility in Support of Operational Hydrology and Water Management" (H-SAF, http://hsaf. meteoam.it/), established by the EUMETSAT Council in 2005, generates and archives high-quality rainfall, soil moisture and snow products for operational hydrological applications, starting from the acquisition and processing of data from Earth observation satellites in geostationary and polar orbits operated both by EUMETSAT and other satellite organizations.

Such an overabundance of products has highly increased the number of studies concerning the assimilation of SM into hydrological models (the reader is referred to [21] for a complete review of these studies). Although many of them show very contrasting results in terms of how the assimilation of SM can improve the accuracy in flood forecasting, all seem to agree that such observations have a high potential to reduce uncertainty in flood prediction. The problem is more related to the preprocessing steps prior to the inclusion of the observations into the hydrological model [22] and in the correct use of the assimilation technique [23], rather than in the value that the observations themselves can bring to flood forecasting. At an operational level, especially in small–medium catchments, the problem is even more exacerbated, due to: (i) the high level of expertise required for implementing and setting up an appropriate assimilation scheme with the risk of not even exploiting these new source of data; and (ii) the lack of uninterrupted and good quality rainfall and evapotranspiration data for running continuous hydrological models, which are needed for current data assimilation approaches.

Recently, [24] proposed a "simplified continuous rainfall runoff model" (SCRRM) that, instead of modelling SM from continuous precipitation and evapotranspiration data, like in classical continuous RR models [12,25,26,27], directly uses SM recorded from external sources (e.g., ground observations, satellite sensors and land surface models) to set the initial conditions of an event-based model. SCRRM explicitly embeds the relationship existing between SM and the model's initial conditions [2,3,4,5] into an event-based RR model to simulate discharge hydrographs. The model was successfully applied in a small catchment of the Attica Region in Greece using different globally-available SM and yielding performances similar to a continuous model [24]. The main advantages of this approach are that SCRRM: (i) does not require continuously-recorded datasets; hence, it can be used also in poorly-gauged areas; (ii) is simple and parsimonious, which is an advantage for users with low hydrological expertise; (iii) requires low computational demand to be run, which makes it appropriate for early warning system applications operating in near real time; and (iv) can be used to "hydro-validate" satellite soil moisture observations. For these reasons, the model is very attractive for civil protection activities, especially in areas where a flood forecasting system is totally absent, and for testing the information content related to the satellite SM dataset.

Based on that, the objectives of this study are two-fold. The first goal is to investigate the performance of two of the satellite SM products of the H-SAF project (*i.e.*, SM-OBS-1 and SM-DAS-2) in order to highlight the information content that they retain in flood modelling. The second goal is to test the performance of SCRRM against the "Modello Idrologico Semi-Distribuito in continuo" (MISDc, [12]) over the Italian territory in order to gain some understanding regarding under what conditions the information derived from the external source of SM can be used for flood forecasting at an operational level.

A total of 593 flood events are used, extracted from a dataset of rainfall and discharges recorded from 2010 to 2013 of 35 Italian catchments, representing a range of sizes, micro-climates, precipitation, streamflows and SM conditions. Through the analysis of such a large number of catchments, some understanding may be gained regarding under what conditions the information derived from the external source of SM can be used for flood forecasting at an operational level.

The paper is organized as follows: In Section 2, we describe the hydrological models used in this study and the characteristics of the selected catchments. In Section 3, we provide a comprehensive analysis of SCRRM performances running with SM-OBS-1 and SM-DAS-2, comparing it against the "Modello Idrologico Semi-Distribuito in continuo" (MISDc, [12]) in terms of reproducing the discharge hydrograph, the peak discharge and the total runoff volume. Finally, inSection 4, we provide the conclusions.

DATA AND METHODS

Catchment Selection and Hydro-Meteorological Data

An available dataset (2010–2013) of 35 Italian catchments having hourly streamflow, precipitation and temperature observations provided by the Italian Civil Protection Department (DPC) was used in the present analysis (see Table 1 for more details). The catchments range in size from 800 to 7400 km^2 (mean size 2507 km2). Most catchments are located in the northern part of Italy (see Figure 1). This region is characterized by a temperate Mediterranean climate with a wet winter and moderate summer rainfall, whereas the south part of Italy has a drier climate with semi-arid summers and temperate winters. The number of catchments is the result of an appropriate selection for excluding catchments with streamflow subject to regulation or diversion and characterized by a number of events less than three. The mean areal rainfall for each catchment was calculated by the GRISO model [28]. GRISO is an improved Kriging-based technique that preserves the values observed at the rain gauge location, allowing for a dynamical definition of the covariance structure associated with each rain gauge by the interpolation procedure. Each correlation structure may depend both on the rain gauge location and on the accumulation time considered. Rainfall events were extracted by selecting those with a continuous rainfall characterized by a total cumulated precipitation larger than 10 mm and no rainfall in the preceding day for a total of 593 flood events. Direct runoff was evaluated as in [29] by using an appropriate base flow separation technique. Mean temperature data for the catchments were obtained by averaging the temperature recorded by the thermometers inside the catchment boundaries.

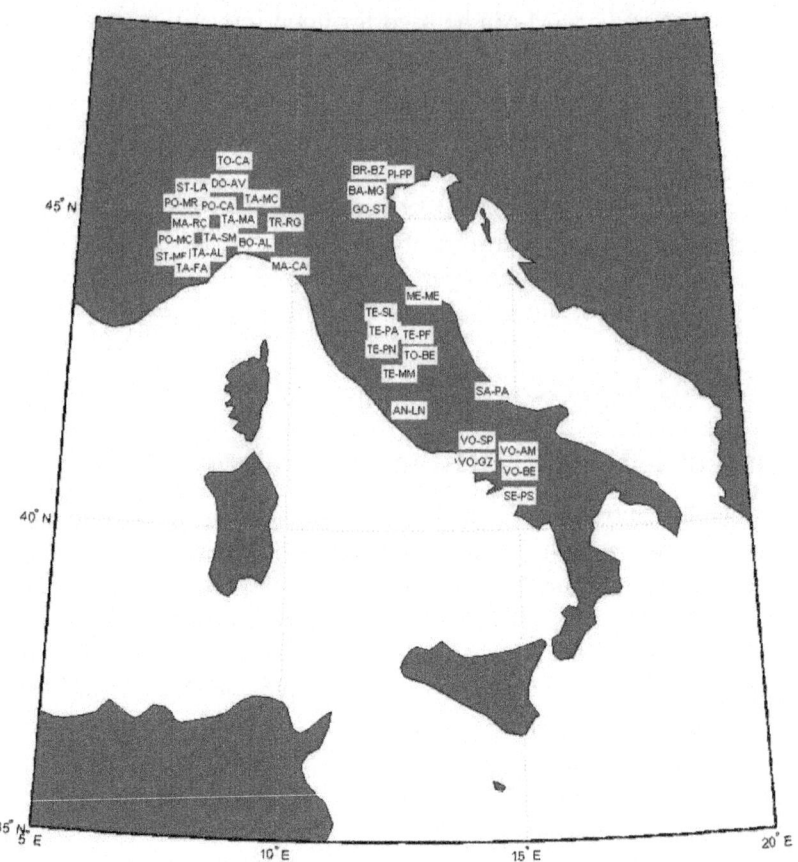

Figure 1. Approximative catchment position.

Table 1. Characteristics of the selected catchments.

#	Code	Catchment	Longi-tude E	Lati-tude N	Area (km2)	No. Events	Mean Annual Rainfall (mm)	Tempera-ture (∘C)
1	AN-LN	Aniene at Lunghezza	12.66	41.93	984.6	5	1525.9	13
2	BA-MG	Bacchiglione at Montegalda	11.67	45.44	1321.3	18	2759.5	10.2
3	BO-AL	Bormida at Alessandria	8.65	44.91	2355.8	18	1157.6	11.6
4	BR-BZ	Brenta at Ber-zizza	11.73	45.78	1506.3	8	2255.1	7

5	DO-AV	Dorabaltea at Verolengo	8.04	45.19	3640.8	11	1089.7	4.4
6	GO-ST	Gorzone at Stanghella	11.76	45.15	1205.8	5	1901.5	13.6
7	MA-CA	Magra at Cal-amazza	9.95	44.2	857.8	26	2807.7	11.3
8	MA-RC	Maira at Raco-nigi	7.67	44.77	967.8	8	998.2	7.9
9	ME-ME	Metauro at Metauro	12.97	43.76	1206.4	13	1617.9	12.6
10	PI-PP	Piave at Ponte-dipiave	12.45	45.71	3902.7	11	2693.3	6.8
11	PO-CA	Po at Carignano	7.69	44.91	3569.5	9	993.6	8.7
12	PO-MC	Po at Moncali-eri	7.68	45	4624.1	9	978.9	9.7
13	PO-MR	Po at Torino Murazzi	7.7	45.06	4899.9	9	986.3	9.7
14	SA-PA	Sangro at Paglieta	14.51	42.21	1522.8	4	1203.6	10.1
15	SE-PS	Sele at Persano Sele	15.03	40.54	2057.9	18	1556.9	12.5
16	ST-LA	Stura di Lanzo at Torino	7.71	45.11	799.9	26	1318.9	7.3
17	ST-MF	Stura di Demonte at Fossano	7.72	44.52	1129.7	8	1300.7	6.3
18	TA-AL	Tanaro at Alba	8.03	44.71	3070.3	20	1170.1	8.8
19	TA-FA	Tanaro at Fari-gliano	7.9	44.52	1364.5	19	1190.2	9.3
20	TA-MA	Tanaro at Masio	8.41	44.87	4157.4	15	1085.9	9.8
21	TA-MC	Tanaro at Mon-tecastello	8.68	44.95	7400	14	1072.5	10.7
22	TA-SM	Tanaro at Asti San Martino	8.21	44.88	3229.7	14	1154.6	9
23	TE-MM	Tevere at Mon-temolino	12.39	42.79	4815.4	26	1341.7	12.8
24	TE-PA	Tevere at Pier-antonio	12.38	43.26	1694.3	12	1397.7	12.2
25	TE-PF	Tevere at Pon-tefelcino	12.43	43.13	1879	28	1395.8	12.3
26	TE-PN	Tevere at Pon-tenuovo	12.43	43.01	3695.3	22	1379.8	12.5

27	TE-SL	Tevere at Santa Lucia	12.24	43.42	837.9	29	1456.9	11.8
28	TO-BE	Topino at Bettona	12.54	43.02	1054.9	24	1333.5	12.6
29	TO-CA	Toce at Candoglia	8.42	45.97	1264	17	1687.9	5.6
30	TR-RG	Trebbia at Rivergaro	9.58	44.9	839.5	27	1735.3	10
31	VO-AM	Volturno at Amorosi	14.45	41.2	1766.8	32	1574.4	13
32	VO-BE	Volturno at Benevento	14.77	41.13	1776.3	34	1173	13.3
33	VO-CA	Volturno at Cancello Arnone	14.02	41.07	4877.9	9	1430.4	13.4
34	VO-GZ	Volturno at Grazzanise	14.11	41.09	4871.1	16	1429.8	13.4
35	VO-SP	Volturno at Solopaca	14.57	41.21	2578.8	29	1307.2	13.3

Satellite and Modelled Soil Moisture Data

In this study, two different SM products, distributed by the H-SAF project, were used covering the period 2010–2013. The products are SM-OBS-1 (large-scale surface soil moisture by radar scatterometer) and SM-DAS-2 (scatterometer data assimilation in the European Centre for Medium-Range Weather Forecasts, ECMWF, Land Data Assimilation System). Given the different catchment extensions and spatial resolution of the two products, only data falling inside the catchment boundaries were selected (when present). When no pixel was contained inside the boundaries, the closest point to the centroid of the catchment was selected. Pixels near the sea or in very high mountain areas were not considered. Table 2summarizes the main characteristics of the selected SM indicators, which are described in detail next.

Table 2. The main characteristics of the soil moisture products used in in this study. SM, soil moisture; ASCAT, Advanced Scatterometer.

Product	Code	Spatial Resolution (km)	Temporal Resolution (days)	Depth (cm)	Source
SM-OBS-1	H07	25	≈ 1	0–2	ASCAT
SM-DAS-2	H14	25	1	0–289	Assimilation of SM-OBS-1

SM-OBS-1

Product SM-OBS-1 is based on the radar scatterometer ASCAT onboard the MetOpsatellites. The instrument scans the scene in a push-broom mode by six side-looking antennas, three left-handed and three right-handed. ASCAT measures radar backscatter at the C-band (5.255 GHz) in VV polarization. The basic instrument sampling distance is 12.5 km. The primary ASCAT observation, sea-surface wind, is processed at 50-km resolution. For SM, processing is performed at 50-km (operational) and 25-km (research) resolution. Global coverage over Europe is achieved in ~1.5 days, while in Italy, measurements are available about once a day. The surface SM product (equivalent to a depth of 0–2 cm of soil) is calculated from the backscatter measurements through a time series-based change detection approach previously used for the ERS-1/2 by [30]. The SM is derived by selecting the historical lowest and highest backscatter measurement to which a 0% (dry) and a 100% (wet) reference is assigned, respectively.

SM-OBS-1 provides knowledge of SM for a very thin surface layer (about 0–2 cm); however, in RR transformation processes, this information may not be sufficient, since the root-zone SM has been shown to be much more important in determining the catchment response to a given storm event [31]. To obtain the root-zone SM product (SWI, Soil Water Index) from the satellite-based surface observations, the recursive formulation [32] of the exponential filter of [30] was adopted:

$$SWI(t_n) = SWI(t_{n-1}) + K_n \left[ms(t_n) - SWI(t_n - 1) \right] \qquad (1)$$

where $ms(t_n)$ is the surface SM observed by the satellite sensor SM-OBS-1, SWI_{tn} is the Soil Wetness Index representing the profile averaged saturation degree and time t_n is the acquisition time of $ms(t_n)$. The gain K_n at time t_n is given by (in a recursive form):

$$K_n = \frac{K_{n-1}}{K_{n-1} + e^{-\left(\frac{t_n - t_{n-1}}{T} \right)}} \qquad (2)$$

where T is the characteristic time length expressed in days and represents the time scale of SM variation to obtain the SWI. For the initialization of this filter, K_1 and SWI_1 were set to 1 and $ms(t_1)$, respectively.

SM-DAS-2

SM-DAS-2 is the first global product of consistent surface and root zone SM that is available in near real time for the numerical weather prediction and climate and hydrological communities. It is based on ASCAT surface SM data assimilation in the ECMWF Land Data Assimilation System. Overall, SM-

DAS-2 relies on an advanced land data assimilation system, which is based on an extended Kalman filter (EKF, [33]) able to ingest information contained in observations close to the surface (2-m temperature and relative humidity synoptic report), as well as new types of data, such as remotely-sensed surface SM. Within the EKF, the surface observation from ASCAT is propagated towards the root region down to 289 cm below the surface, providing estimates for 4 layers (thicknesses of 7, 21, 72 and 189 cm). The ECMWF model generates SM profile information according to the Hydrology Tiled ECMWF Scheme for Surface Exchanges over Land (HTESSEL, [19]). SM-DAS-2 is available at a 24-hour time step with a spatial resolution of 25 km, with a global daily coverage at 00:00 UTC. SM-DAS-2 is run continuously in order to ensure the time series consistency of the product (and also to provide values when there is no satellite data, from the model propagation). SM-DAS-2 SM has been validated against *in situ* SM from many locations in Africa, Australia and Europe, providing satisfactory results [34]. It has to be noted that SM-DAS-2 data prior to 2012 are experimental data (*i.e.*, the full 2010–2013 dataset might not be completely consistent), so it is highly expected that it can provide better performances from 2012 onwards.

In this study, the information of the SM for any soil depth between 0 and 289 cm, $\theta_{SM}-DAS_2$, was obtained by the weighed mean of the SM provided by the related layer, according to:

$$\theta_{SM-DAS2} = \theta_1 \qquad\qquad : z \le 7\text{cm}$$

$$\theta_{SM-DAS2} = \frac{(\theta_1 7 + \theta_2(z - 7))}{z} \qquad\qquad : 7 < z \le 28\text{cm}$$

$$\theta_{SM-DAS2} = \frac{(\theta_1 7 + \theta_2 21 + \theta_3(z - 28))}{z} \qquad\qquad : 28 < z \le 100\text{cm}$$

$$\theta_{SM-DAS2} = \frac{(\theta_1 7 + \theta_2 21 + \theta_2 72 + \theta_3(z - 100))}{z} \qquad\qquad : 100 < z \le 289\text{cm} \tag{3}$$

where $\theta_1, \theta_2, \theta_3, \theta_4$ are the SM for each of the four layers of the SM-DAS-2 product and z is the parameter representing the depth.

Hydrological Models

In the following, we present a description of all of the hydrological models used for this study. A continuous model (MISDc, "Modello Idrologico Semiditribuito in Continuo", [12]) is used as a benchmark for evaluating the reliability of the SCRRM model presented in Section 2.3.2 running with the two SM products of the H-SAF project described in Section 2.2.

Continuous Model: MISDc

MISDc [12] is a continuous rainfall runoff model successfully applied to the Tiber River for flood prediction and operational purposes. The lumped version of the MISDc model used in this study couples a soil water balance model (SWB, [2]) to simulate the SM temporal pattern and a routing module [35] for transferring the rainfall excess to the outlet section of the catchment. The two models are linked through an experimentally-derived linear relationship between the potential maximum soil moisture retention S of the Soil Conservation Service-Curve Number (SCS-CN, [36]) and the relative SM at the beginning of the event.

The SWB model is a simple water balance model representing the main processes needed for SM simulation: infiltration, percolation and evapotranspiration. The processes are represented for infiltration through the Green–Ampt equation for drainage by a gravity-driven non-linear relationship and for actual evapotranspiration by a linear relationship with the potential evapotranspiration, calculated through a modified Blaney and Criddle method. The reader is referred to [2] and [12] for a detailed description of the model. The SM simulated by the SWB is used to calculate the parameter S method by means of an experimentally-derived relationship between S and SM [10]:

$$S = a(1 - \theta_e)$$

(4)

where θ is the modelled relative SM at the beginning of the event and a is a parameter to be estimated. Once the S parameter is estimated, the routing to the outlet of the catchment is obtained from the convolution of the rainfall excess and the geomorphological instantaneous unit hydrograph (GIUH), such as proposed by [37]. In the model, the lag time is evaluated through the relationship proposed by [29]:

$$L = \eta 1.19 A^{0.33}$$

(5)

with L being the lag time in hours, A the area of the catchment (km^2), and η the parameter to be calibrated [38].

MISDc requires as input data the rainfall and air temperature. The model outputs are both the direct runoff in correspondence to selected flood events and the catchment average relative SM.

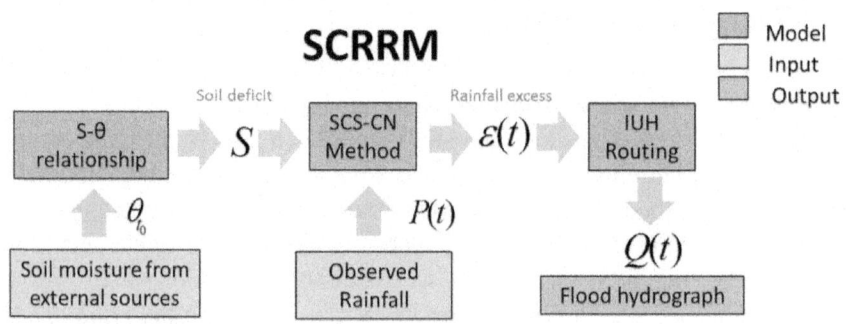

Figure 2. Structure of the simplified continuous rainfall-runoff model (SCRRM).

SCRRM

Unlike MISDc, where the SM at the beginning of the event is simulated by the SWB model, in SCRRM (Figure 2), it is provided by an external indicator, *i.e.*, from satellite SM observations or model-based reanalysis products. SCRRM reflects the structure of MISDc, but has some significant differences. Indeed, the temporal evolution of the soil wetness conditions of the catchment is not modelled from rainfall and temperature data, as in MISDc, but it is integrated directly into the model from SM observations (*i.e.*, SWB is replaced in SCRRM by SM observations immediately before the rainfall event) considering the SM products as proxies for the assessment of the wetness state of the catchment. The model can be also seen as the result of an assimilation technique in which the SM observations from the external source are assumed to be error free (*i.e.*, a direct insertion technique).

Like in MISDc, the model exploits the observed linear behaviour between the wetness state of the soil and the parameter S of the SCS method by Equation (4). Once S is known, the rainfall excess is calculated by:

$$\epsilon = \frac{(P - F_a)^2}{P - F_a + S} \text{ if } \epsilon \geq F_a$$

(6)

where F_a is the initial abstraction, ϵ is the rainfall excess and P is the rainfall depth. As in the SCS method, the quantity F_a is considered linearly dependent on S by:

$$F_a = \lambda S$$

(7)

and then is transferred to the outlet section with the same routing module used by MISDc. In Equation (7), λ is the initial abstraction coefficient.

In synthesis, SCRRM uses the SM and the event rainfall data as the sole inputs to simulate hourly flood hydrographs. Since the SM is provided by an external indicator, Equation (4) becomes a model relation embedded in the model structure. The calibration of the model involves the following three parameters: the coefficient of initial abstractions λ, the parameter a of the S$-\theta$ relationship and the parameter η of Equation (5).

As remarked in Section 2.2, the soil layer depth that controls the RR transformation is usually larger than a few centimetres. As a result, the application of SCRRM with the H-SAF products is taken into account by including the soil depth as an additional parameter of SCRRM. For SM-OBS-1, such a parameter is controlled by the characteristic time length T of the exponential filter described in Section 2.2.1, while for SM-DAS-2, the parameter z of Equation (3) was considered, which can vary from 0 to 289 cm (see Section 2.2.2). For this study, a lumped model was employed, even though the same concept can be easily applied to spatially-distributed models.

Performance Indexes

The performances of SCRRM and MISDc were evaluated by considering different indexes. The first one, commonly used for assessing the agreement between simulated and observed hydrographs, is the Nash–Sutcliffe efficiency coefficient, NS[39]:

$$NS = 1 - \frac{\sum_{t=1}^{T_{ev}} (Q_{obs} - Q_{sim})^2}{\sum_{t=1}^{T_{ev}} (Q_{obs} - \bar{Q}_{obs})^2} \tag{8}$$

where Q_{obs} and Q_{sim} are the observed and simulated discharges at time t, respectively, $\bar{Q}_{obs}$ is the mean value of the observed discharge during the event and T_{ev} is the event duration. NS was calculated for each of the selected events for every catchment considered in the analysis. In particular, the mean NS calculated over the selected events for each catchment was used as an objective function for calibrating the parameters of the models:

$$\overline{NS} = \frac{\sum_{j=1}^{N_{ev}} NS_j}{N_{ev}} \tag{9}$$

where N_{ev} is the number of events considered, whereas index j varies between 1 and the number of selected events for each catchment. For model calibration, a standard gradient-based automatic optimisation method (the "fmincon" function in MATLAB ®,[40]) was used. In addition, to evaluate the performance of the model in reproducing flood events, the percentage error on peak discharge:

$$E_{Q_p} = \frac{\max(Q_{obs}) - \max(Q_{sim})}{\max(Q_{obs})} \tag{10}$$

and the percentage error on direct runoff volume:

$$E_V = \frac{\sum\limits_{t}^{T_{ev}} Q_{obs} - \sum\limits_{t}^{T_{ev}} Q_{sim}}{\sum\limits_{t}^{T_{ev}} Q_{obs}} \tag{11}$$

were both evaluated for each single event and as the mean of all of the selected events for every catchment considered in this study.

RESULTS AND DISCUSSION

In the following, we show the performances obtained by SCRRM using SM-OBS-1 and SM-DAS-2 compared with the results obtained by MISDc. Both MISDc and SCRRM were calibrated using Equation (9) as the objective function and tuning the four parameters of SCRRM and the seven parameters of MISDc. The calibration was carried out separately for each catchment yielding $\overline{NS}$ equal to 0.61, 0.58 and 0.66 for SM-OBS-1, SM-DAS-2 and MISDc, respectively.

In Figure 3(a) and Figure 3(b), the histograms of z and T (*i.e.*, the parameters representing the influence of the soil depth in the RR transformation) show higher frequencies around 1–90 days for T and two main clusters for z: one below 20 cm and the other above 105 cm. The median T and z are about 48 days and 1 m, respectively, which are quite acceptable values in Italy (e.g., [31]). Values of T above 100 days are obtained for catchments PI-PP (Piave at Pontedipiave), AN-LN (Aniene at Lunghezza), BR-BZ (Brenta at Berzizza) and DO-AV (Dorabaltea at Verolengo). These values could be due to several reasons: (i) those related to the influence of a deeper soil layer in the RR transformation (catchments PI-PP and DO-AV have both $T = 300$ days and $z = 80$ cm and 130 cm, which are quite consistent values); (ii) and those related to the accuracy of the observations (e.g., observed discharge datasets), which may affect the consistency of the calibration procedure. Finally, another important reason could be the accuracy of the satellite observations, which results in contrasting T values if compared with z (z = 6 cm and T = 256 days for catchment BR-BZ).

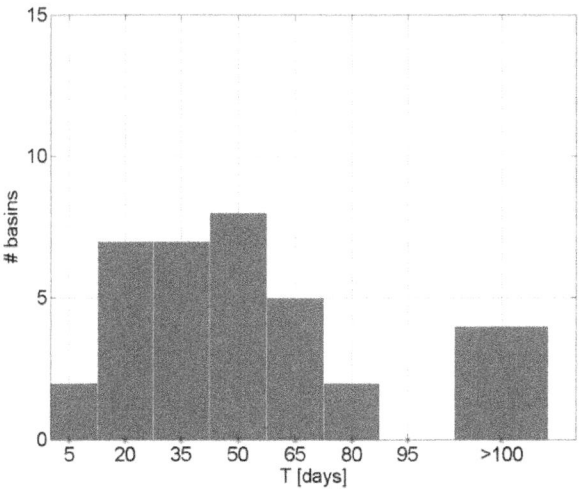

(a) Characteristic time length T

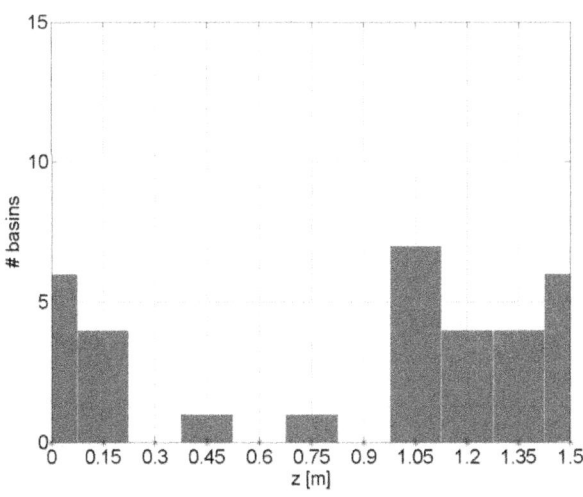

(b) Soil depth z

Figure 3. Histogram of the parameters of SCRRM representing the soil depth involved in the rainfall-runoff transformation. T = characteristic time length used in SCRRM with SM-OBS-1 (**a**); z = soil depth used in SCRRM with SM-DAS-2 (**b**).

Figure 4(a) shows the median NS50 obtained by SM-OBS-1, SM-DAS-2 and MISDc for the selected catchments. As can be seen, there is a general agreement between the three configurations, with MISDc being generally better than SM-OBS-1 and SM-DAS-2. The results for E_{Qp50} and E_{V50} in Figure 4(b) and Figure 4(c) reflect those for NS, with the line representing MISDc generally lower than the lines of SM-OBS-1 and SM-DAS-2.

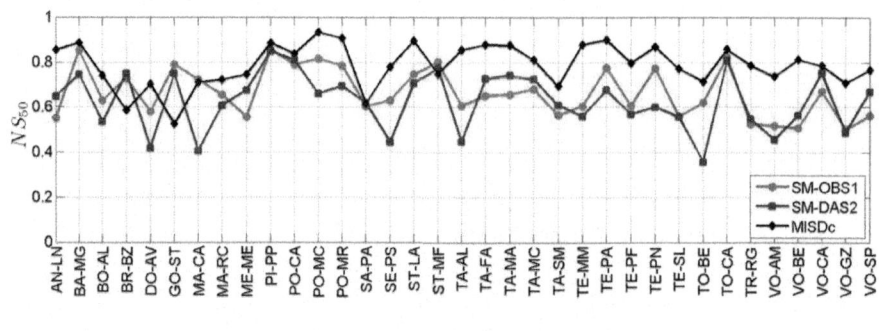

(a) NS_50

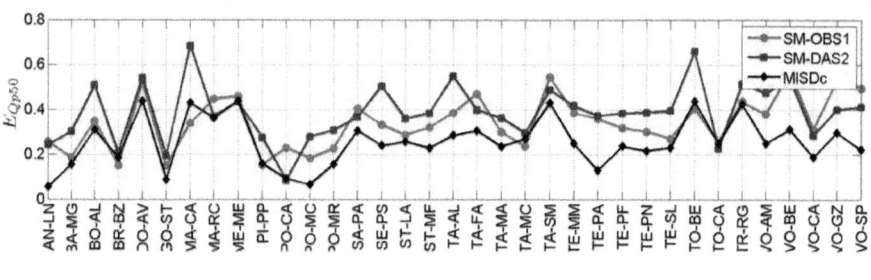

(b) E_{Qp50}

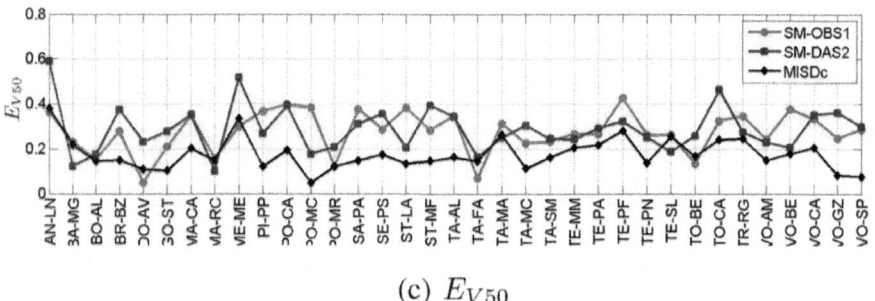

(c) E_{V50}

Figure 4. Median performance indexes for the selected catchments, NS50 =median Nash–Sutcliffe (**a**); EQp50 = median relative error in peak discharge (**b**); EV50 = median relative error in volume (**c**).

The reasons for this could be attributed to: (i) the better description of the water balance that MISDc can provide with respect to the SCRRM (indeed, MISDc uses the water balance model to "adjust" the initial condition prior to the RR event); (ii) the underlying errors present in the SM estimates and in the coarser resolution of the H-SAF SM products compared to the processes happening at the catchment scale; and (iii) the larger number of parameters of MISDc with respect to SCRRM. It is worth noting that the sensitivity of the selected catchments to the initial conditions was tested by running the SCRRM model with a constant value of soil moisture (*i.e.*, equal to the mean of the soil moisture values associated with the events selected for each catchment). For both products, a sensible reduction of 15% was found for mean NS and an increase of 6% for both E_{Qp} and E_v.

The performance of SCRRM with respect to MISDc can be better visualized by the scatter plots shown in Figure 5(a) and Figure 5(b). The figures plot the performance indexes NS_{50}, E_{Qp50} and E_{V50} of MISDc against those of SM-OBS-1 and SM-DAS-2, respectively. Overall, MISDc outperforms SCRRM, but it can be recognized that when MISDc provides a good performance, also SCRRM works well; thus, we may conclude that the results are somehow dependent on the quality of the input data and on the model structure.

Overall, the median of NS, calculated on all selected events, yields 0.65, 0.60 and 0.74, for SM-OBS-1, SM-DAS-2 and MISDc, respectively, while 0.35, 0.40 and 0.25 are obtained for E_{Qp}. Similarly, E_v provides 0.29, 0.30 and 0.16. Note that MISDc generally outperforms SCRRM in both cases of SM-OBS-1 and SM-DAS-2, except for some catchments (e.g., for NScatchments MA-CA (Magra at Calamazza), GO-ST (Gorzone at Stanghella), ST-MF (Stura di Demonte at Fossano) and BR-BZ; see 4(a); for E_v, also MA-RC (Maira at Raconigi), TA-FA (Tanaro at Farigliano) and TO-BE (Topino at Bettona); seeFigure 4(c)). Since the structures of MISDc and SCRRM are very similar (*i.e.*, they rely on the same runoff production and routing models), the only reason for explaining the higher performance is the better estimation of SM at the beginning of the flood event. Indeed, in this study, MISDc is run in lumped mode, providing an estimate of SM for the entire catchment. On the contrary, the SM time series used in SCRRM rely on SM observations from different parts of the basin area; thus, it is likely that the spatial variability of SM is better taken into account by SCRRM. Moreover, SM estimates by MISDc rely on rainfall and temperature measurements, which, other than possibly being affected by many error sources [41], are averaged throughout the basin area, producing possible additional uncertainties [13].

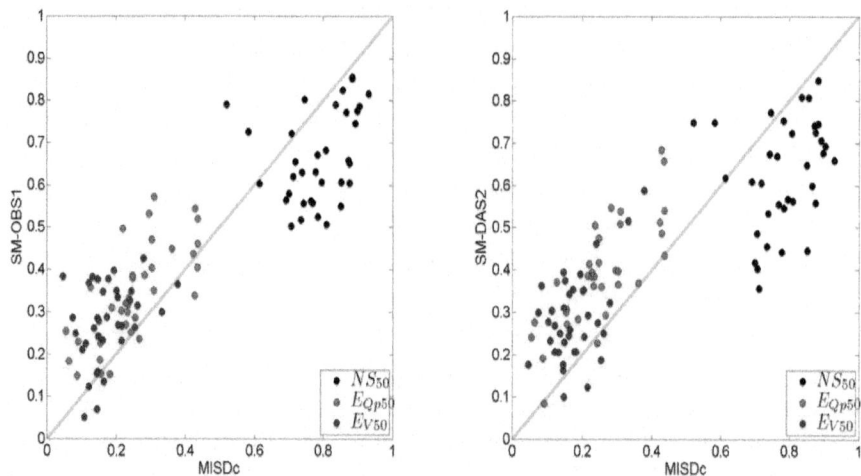

Figure 5. Comparison between MISDc (Modello Idrologico Semi-Distribuito in continuo) and SCRRM run with SM-OBS-1 (**a**) and SM-DAS-2 (**b**) in terms of median Nash–Sutcliffe, NS_{50}, median relative error on peak discharge, E_{Qp50}, and median relative error in total runoff volume, E_{V50}, for the selected catchments.

To assess if SCRRM results could be affected by catchment area extension, climatic conditions, average temperature and latitude, we calculated $\overline{NS}$ against these variables (not shown for the sake of brevity), but we did not find any strong evidence of any existing trend, except a slight increment of the performances as the area of the catchment is increased. Although we cannot draw any conclusions, because of the relatively small number of catchments analyzed (and the possible errors contained in the observed discharge), this increment is highly expected due to the coarse resolution of SM-OBS-1 and SM-DAS-2. Future investigations will consider a longer dataset and stronger climatic differences for the catchments in order to gain additional information on the SCRRM performances.

Finally, in Figure 6(a) and Figure 7(a), we plotted the simulated and the observed discharge for two representative catchments (TE-SL (Tevere at Santa Lucia) and PO-CA (Po at Carignano)). The respective SM time series in terms of SWI, $\theta_{SM-DAS-2}$ and SM simulated by MISDc are plotted in Figure 6(b) and Figure 7(b). Table 3 summarizes the results obtained for the two catchments.

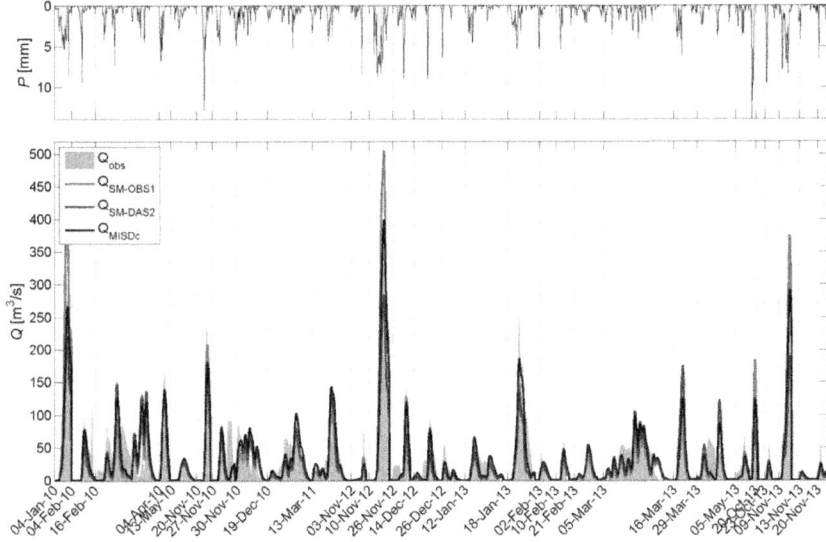

(a) Simulated *vs.* observed discharge for the selected events.

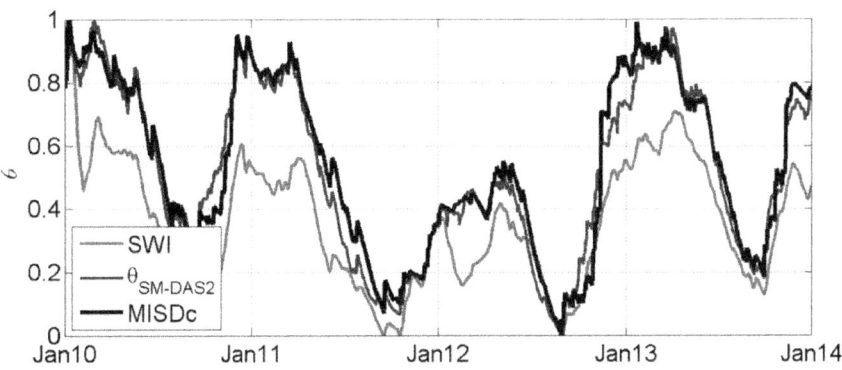

(b) Averaged SWI, $\theta_{SM-DAS-2}$ and MISDc soil moisture.

Figure 6. Observed rainfall (P) and simulated *versus* observed discharge (Q) for the selected events (**a**). Averaged SWI (Soil Water Index)(Equation (1)), $\theta_{SM-DAS-2}$(Equation (3)) and MISDc soil moisture, , from 2010 to 2013 (**b**). Catchment: Tevere River at Santa Lucia (TE-SL).

For TE-SL (with an area of about 840 km²), the models behave very similar, with MISDc slightly superior to SCRRM for all of the considered performance scores. In particular, SM-OBS-1 tends to overestimate the direct runoff volume and the peak discharges, both in the mean and in the median, especially for the largest events (see Figure 6(a)), providing the largest values of both E_{Qp} and E_V (see 3). The SM time series show very similar patterns for MISDc and $\theta_{SM-DAS-2}$ ($z = 166$ cm), while SWI seems characterized by a lower variability ($T = 48$ days).

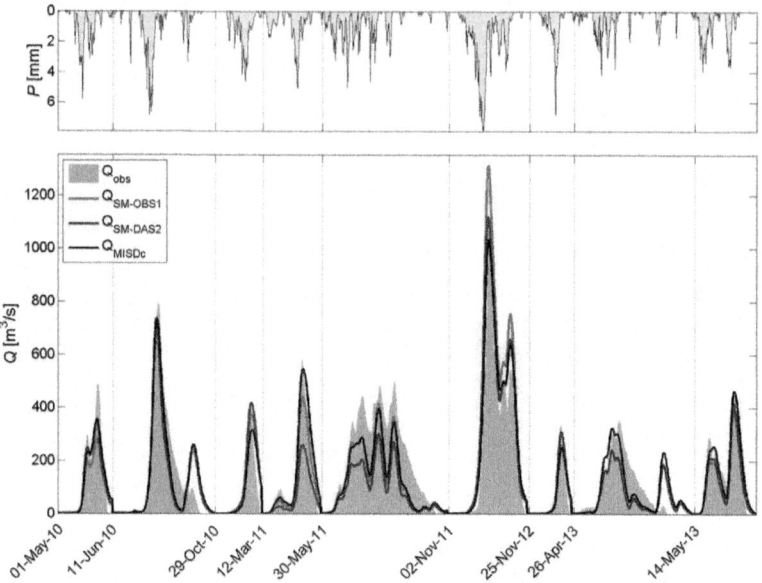

(a) Simulated *vs.* observed discharge for the selected events.

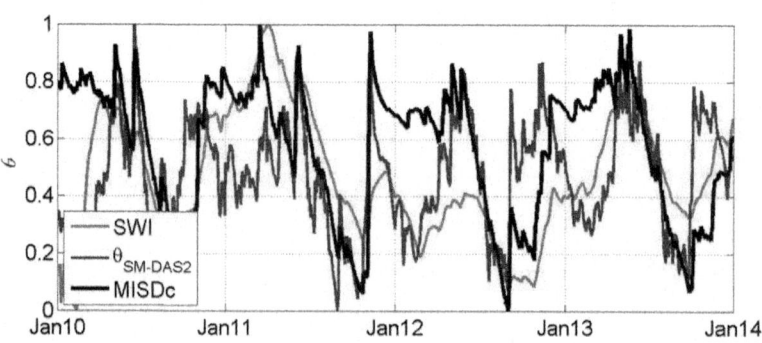

(b) Averaged SWI, $\theta_{SM-DAS-2}$ and MISDc soil moisture.

Figure 7. Observed rainfall (P) and simulated *versus* observed discharge (Q) (**a**). Averaged SWI (Equation (1)), $\theta_{SM-DAS-2}$ (Equation (3)) and MISDc soil moisture, θ, from 2010 to 2013 (**b**). Catchment: Po River at Carignano (PO-CA).

For PO-CA (with an area of about 3570 km2, four-times the size of TE-SL), all of the models provide good results, both in the mean and in the median scores. A general agreement between the models can be also seen in Figure 6(a). It is interesting to note the good performance obtained by SM-DAS-2, which provides the lowest value of the median E_v. In contrast with TE-SL, the SM time series for PO-CA in 7(b) appear very different with respect to MISDc. Indeed, SWI ($T = 82$ days) and $\theta_{SM-DAS-2}$ ($z = 168$ cm) are the results of the average of multiple pixels falling inside the catchment boundaries of PO-CA; thus, the spatial variability and the inherent errors present in the observed time series may significantly affect the value of the calibrated parameters, as well as the averaged SM temporal pattern throughout the basin.

Table 3. Summary of the results obtained for the catchments Tevere River at Santa Lucia (TE-SL) and Po River at Carignano (PO-CA).

Catchment	Model		Mean			Median	
		NS	EQp	EV	NS	EQp	EV
	SM-OBS-1	0.452	0.298	0.401	0.558	0.269	0.367
TE-SL	SM-DAS-2	0.427	0.368	0.354	0.555	0.394	0.268
	MISDc	0.556	0.259	0.180	0.770	0.228	0.121
	SM-OBS-1	0.744	0.223	0.253	0.790	0.229	0.210
PO-CA	SM-DAS-2	0.692	0.213	0.305	0.808	0.084	0.277
	MISDc	0.805	0.135	0.146	0.836	0.090	0.102

CONCLUSIONS

In this paper, a simplified continuous rainfall-runoff model using SM from external sources for initialization has been used for testing the performance of two operational products (SM-OBS-1 and SM-DAS-2) of the "EUMETSAT Satellite Application Facility in Support of Operational Hydrology and Water Management (H-SAF)" and to gain some knowledge regarding under what conditions SCRRM can be used for operational purposes. The model was applied in 35 catchments of the Italian territory covering different sizes and climate conditions. The performances obtained by SCRRM in flood modelling have been compared against those of a classical continuous model (MISDc, [12]) showing satisfactory results (mean NS equal to 0.61, 0.58 for SM-OBS-1 and SM-DAS-2, respectively), but generally lower than MISDc (mean NS=0.66).

In particular, it was found that:

- In 35 Italian catchments (800 to 7400 km2), satisfactory results can be obtained by the use SM-OBS-1 and SM-DAS-2 within SCRRM providing a median NS calculated on all of the selected events equal to 0.65 and 0.60, respectively. Similarly, the relative errors in median peak discharge and in runoff volume are 0.35 and 0.29 (SM-OBS-1) and 0.40 and 0.30 (SM-DAS-2). This means that the two products: (i) provide very similar performance and may both be satisfactorily used within SCRRM; and (ii) offer similar information content in flood modelling, which can be efficiently exploited in the context of soil moisture data assimilation in continuous models.

- MISDc generally outperforms SCRRM, except in a few cases. Although this aspect needs further investigation, the reason could be due to the fact that MISDc was run in lumped mode, while SCRRM SM was obtained by averaging the value of SM of different pixels falling inside the catchment boundaries, thus taking more into account for the SM spatial variability. In any case, although the performances of SCRRM are generally lower than those of MISDc (but not by far), they highlight two main interesting issues. First, for operational purposes, SCRRM is expected to be a valuable alternative to a continuous model in poorly gauged areas, since its structure is less sensitive to problems (not rare) of rain-gauge malfunctions and breakage. Second, the satisfactory results obtained indicate that H-SAF soil moisture products have the potential to improve flood modelling if used with more complex data assimilation schemes with continuous models (*i.e.*, by assimilating SM-OBS-1 and SM-DAS-2 products into the MISDc model).

- Median T and z values, (*i.e.*, the parameters representing the influence of the soil depth in the RR transformation) are 48 days and 100 cm, respectively, which are quite reasonable values in Italy.

- SCRRM can be used as a "hydro-validation tool" to assess the performance of different soil moisture products in terms of the ability to reproduce flood hydrographs. This is a new method for validating soil moisture data that has not been used before.

Overall, we can conclude that both products, SM-OBS-1 and SM-DAS-2, contain sufficient information for satisfactorily reproducing floods. The SCRRM benefits are the possibility to be used in poorly gauged areas (e.g., in areas characterized by discontinuous measures of rainfall and temperature), its simplicity and parsimony, which facilitate setup and operational use. On the other hand, it must be said that in areas characterized by dense and robust hydro-meteorological networks, like the ones used in this paper, a continuous modelling approach is preferable, since it allows one to obtain better performance. In this context, the satellite soil moisture products could be optimally integrated into the continuous model for obtaining a better modelling chain for flood forecasting. Future applications of the model will consider the use of satellite rainfall observations in order to rely completely on satellite data.

ACKNOWLEDGMENTS

The authors wish to thank EUMETESAT in the framework of the EUMETSAT Satellite Application Facility in Support of Operational Hydrology and Water Management (H-SAF) project and the Italian Civil Protection Department for providing the relevant datasets used in the paper.

AUTHOR CONTRIBUTIONS

All authors contributed extensively to the work presented in this paper. Specific contributions included the development of the SCRRM and MISDc models (Christian Massari, Luca Brocca), of the two H-SAF soil moisture products SM-OBS-1 (Wolfgang Wagner) and SM-DAS-2 (Clement Albergel, Patricia de Rosnay), acquisition and processing of ground-based hydrometeorological data (Simone Gabellani, Silvia Puca, Luca Ciabatta), and preparation of the manuscript and figures (Christian Massari, Luca Brocca, Luca Ciabatta, Tommaso Moramarco). All co-authors contributed to the editing of the manuscript and to the discussion and interpretation of the results.

REFERENCES

1. Jongman, B.; Hochrainer-Stigler, S.; Feyen, L.; Aerts, J.C.J.H.; Mechler, R.; Botzen, W.J.W.; Bouwer, L.M.; Pflug, G.; Rojas, R.; Ward, P.J. Increasing stress on disaster-risk finance due to large floods. *Nat. Clim. Chang.* **2014**, *4*, 264–268.

2. Brocca, L.; Melone, F.; Moramarco, T. On the estimation of antecedent wetness conditions in rainfall-runoff modelling.*Hydrol. Process.* **2008**, *642*, 629–642.

3. Coustau, M.; Bouvier, C.; Borrell-Estupina, V.; Jourde, H. Flood modelling with a distributed event-based parsimonious rainfall-runoff model: Case of the karstic Lez river catchment. *Nat. Hazard. Earth Syst. Sci.* **2012**, *12*, 1119–1133.

4. Beck, H.E.; Jeu, R.A.M.D.; Schellekens, J.; Dijk, A.I.J.M.V.; Bruijnzeel, L.A. Improving curve number based storm runoff estimates using soil moisture proxies. *IEEE J. Sel. Top. Appl. Earth Obs. Remote Sens.* **2010**, *2*, 250–259.

5. Tramblay, Y.; Bouaicha, R.; Brocca, L.; Dorigo, W.; Bouvier, C.; Camici, S.; Servat, E. Estimation of antecedent wetness conditions for flood

modelling in northern Morocco. *Hydrol. Earth Syst. Sci.* **2012**, *16*, 4375–4386.

6.   Viviroli, D.; Mittelbach, H.; Gurtz, J.; Weingartner, R. Continuous simulation for flood estimation in ungauged mesoscale catchments of Switzerland–Part II: Parameter regionalisation and flood estimation results. *J. Hydrol.* **2009**,*377*, 208–225.

7.   Van Steenbergen, N.; Willems, P. Increasing river flood preparedness by real-time warning based on wetness state conditions. *J. Hydrol.* **2013**, *489*, 227–237.

8.   Aronica, G.; Candela, A. A regional methodology for deriving flood frequency curves (FFC) in partially gauged catchments with uncertain knowledge of soil moisture conditions. In Proceedings of iEMSs, Osnabru, Germany, 14–17 June 2004; pp. 1147–1183.

9.   Brocca, L.; Melone, F.; Moramarco, T.; Morbidelli, R. Antecedent wetness conditions based on ERS scatterometer data.*J. Hydrol.* **2009**, *364*, 73–87.

10.   Brocca, L.; Melone, F.; Moramarco, T.; Singh, V.P. Assimilation of observed soil moisture data in storm rainfall-runoff modeling. *J. Hydrol. Eng.* **2009**, *14*, 153–165.

11.   Tramblay, Y.; Bouvier, C.; Martin, C.; Didon-Lescot, J.F.; Todorovik, D.; Domergue, J.M. Assessment of initial soil moisture conditions for event-based rainfall-runoff modelling. *J. Hydrol.* **2010**, *387*, 176–187.

12.   Brocca, L.; Melone, F.; Moramarco, T. Distributed rainfall-runoff modelling for flood frequency estimation and flood forecasting. *Hydrol. Process.* **2011**, *25*, 2801–2813.

13.   Tramblay, Y.; Bouvier, C.; Ayral, P.A.; Marchandise, A. Impact of rainfall spatial distribution on rainfall-runoff modelling efficiency and initial soil moisture conditions estimation. *Nat. Hazard. Earth Syst. Sci.* **2011**, *11*, 157–170.

14.   Aubert, D.; Loumagne, C.; Oudin, L. Sequential assimilation of soil moisture and streamflow data in a conceptual rainfall-runoff model. *J. Hydrol.* **2003**, *280*, 145–161.

15.   Dorigo, W.A.; Wagner, W.; Hohensinn, R.; Hahn, S.; Paulik, C.; Xaver, A.; Gruber, A.; Drusch, M.; Mecklenburg, S.; van Oevelen, P.; *et al.* The International Soil Moisture Network: A data hosting facility for global in situ soil moisture measurements. *Hydrol. Earth Syst. Sci.* **2011**, *15*, 1675–1698.

16.   Bartalis, Z.; Wagner, W.; Naeimi, V.; Hasenauer, S.; Scipal, K.; Bonekamp, H.; Figa, J.; Anderson, C. Initial soil moisture retrievals from

the METOP-A Advanced Scatterometer (ASCAT). *Geophys. Res. Lett.* **2007**, *34*, L20401.

17. Owe, M.; de Jeu, R.; Holmes, T. Multisensor historical climatology of satellite-derived global land surface moisture. *J. Geophys. Res.* **2008**, *113*, F01002.

18. Kerr, Y.H.; Waldteufel, P.; Wigneron, J.P.; Delwart, S.; Cabot, F.; Boutin, J.; Escorihuela, M.J.; Font, J.; Reul, N.; Gruhier, C.; *et al.* The SMOS mission: New tool for monitoring key elements of the global water cycle. *Proc. IEEE* **2010**,*98*, 666–687.

19. Balsamo, G.; Albergel, C.; Beljaars, A.; Boussetta, S.; Brun, E.; Cloke, H.; Dee, D.; Dutra, E.; Pappenberger, F.; de Rosnay, P.; *et al.* ERA-Interim/Land: A global land-surface reanalysis based on ERA-Interim meteorological forcing.*Era Rep. Series* **2012**, *13*, 1–25.

20. Wagner, W.; Bloschl, G.; Pampaloni, P.; Calvet, J.; Bizzarri, B.; Wigneron, J.; Kerr, Y. Operational readiness of microwave remote sensing of soil moisture for hydrologic applications. *Nordic Hydrol.* **2007**, *38*, 1–20.

21. Liu, Y.; Weerts, A.H.; Clark, M.; Hendricks Franssen, H.J.; Kumar, S.; Moradkhani, H.; Seo, D.J.; Schwanenberg, D.; Smith, P.; van Dijk, A.I.J.M.; *et al.* Advancing data assimilation in operational hydrologic forecasting: Progresses, challenges, and emerging opportunities. *Hydrol. Earth Syst. Sci.* **2012**, *16*, 3863–3887.

22. Reichle, R.H. Bias reduction in short records of satellite soil moisture. *Geophys. Res. Lett.* **2004**, *31*, L19501.

23. Crow, W.T.; Van Loon, E. Impact of incorrect model error assumptions on the sequential assimilation of remotely sensed surface soil moisture. *J. Hydrometeorol.* **2006**, *7*, 421–432.

24. Massari, C.; Brocca, L.; Barbetta, S.; Papathanasiou, C.; Mimikou, M.; Moramarco, T. Using globally available soil moisture indicators for flood modelling in Mediterranean catchments. *Hydrol. Earth Syst. Sci.* **2014**, *18*, 839–853.

25. Todini, E. Rainfall-runoff modeling—Past, present and future. *J. Hydrol.* **1988**, *100*, 341–352.

26. Beven, K.; Freer, J. A dynamic TOPMODEL. *Hydrol. Process.* **2001**, *15*, 1993–2011.

27. Famiglietti, J.S.; Wood, E.F. Multiscale modeling of spatially variable water and energy balance processes. *Water Resour. Res.* **1994**, *30*, 3061–3078.

28. Pignone, F.; Rebora, N.; Silvestro, F.; Castelli, F. *GRISO (Generatore Random di Interpolazioni Spaziali da Osservazioni incerte). Relazione delle attività del I anno inerente la Convenzione 778/2009 tra Dipartimento di Protezione Civile e Fondazione CIMA (Centro Internazionale in Monitoraggio Ambientale)*; Technical Report No. 272; Savona, Italy; p. 353. 2010.

29. Melone, F.; Corradini, C.; Singh, V.P. Lag prediction in ungauged basins: An investigation through actual data of the upper Tiber River valley. *Hydrol. Process.* **2002**, *16*, 1085–1094.

30. Wagner, W.; Lemoine, G.; Rott, H. A method for estimating soil moisture from ERS scatterometer and soil data.*Remote Sens. Environ.* **1999**, *70*, 191–207.

31. Brocca, L.; Melone, F.; Moramarco, T.; Wagner, W.; Naeimi, V.; Bartalis, Z.; Hasenauer, S. Improving runoff prediction through the assimilation of the ASCAT soil moisture product. *Hydrol. Earth Syst. Sci.* **2010**, *14*, 1881–1893.

32. Albergel, C.; Rüdiger, C.; Carrer, D.; Calvet, J.C.; Fritz, N.; Naeimi, V.; Bartalis, Z.; Hasenauer, S. An evaluation of ASCAT surface soil moisture products with *in-situ* observations in Southwestern France. *Hydrol. Earth Syst. Sci.* **2009**,*13*, 115–124.

33. De Rosnay, P.; Drusch, M.; Vasiljevic, D.; Balsamo, G.; Albergel, C.; Isaksen, L. A simplified Extended Kalman Filter for the global operational soil moisture analysis at ECMWF. *Q. J. R. Meteorol. Soc.* **2013**, *139*, 1199–1213.

34. Albergel, C.; de Rosnay, P.; Gruhier, C.; Muñoz Sabater, J.; Hasenauer, S.; Isaksen, L.; Kerr, Y.; Wagner, W. Evaluation of remotely sensed and modelled soil moisture products using global ground-based *in situ* observations. *Remote Sens. Environ.* **2012**, *118*, 215–226.

35. Melone, F.; Neri, N.; Morbidelli, R.; Saltalippi, C. A conceptual model for flood prediction in basins of moderate size. In *Applied simulation and Modeling*; Hamza, M.H., Ed.; IASTED Acta Press: Anaheim, CA, USA, 2001; pp. 461–466.

36. Kim, N.; Lee, J. Temporally weighted average curve number method for daily runoff simulation. *Hydrol. Process.* **2008**,*4948*, 4936–4948.

37. Gupta, V.; Waymire, C. A representation of an instantaneous unit hydrograph from geomorphology. *Water Resour. Res.* **1980**, *16*, 855–862.

38. Moramarco, T.; Melone, F.; Singh, V.P. Assessment of flooding in urbanized ungauged basins: A case study in the Upper Tiber area, Italy. *Hydrol. Process.* **2005**, *19*, 1909–1924.

39. Nash, J.; Sutcliffe, J. River flow forecasting through conceptual models part I—A discussion of principles. *J. Hydrol.***1970**, *10*, 282–290.

40. Bober, W. *Introduction to Numerical and Analytical Methods with MATLAB for Engineers and Scientists*; CRC Press, Inc.: Boca Raton, FL, USA, 2013.

41. Crow, W.T.; van den Berg, M.J.; Huffman, G.J.; Pellarin, T. Correcting rainfall using satellite-based surface soil moisture retrievals: The Soil Moisture Analysis Rainfall Tool (SMART). *Water Resour. Res.* **2011**, *47*, W08521.

CITATION

CHAPTER 1

Mario Lefebvre and Fatima Bensalma, "An Application of Filtered Renewal Processes in Hydrology,"International Journal of Engineering Mathematics, vol. 2014, Article ID 593243, 9 pages, 2014. doi:10.1155/2014/593243

CHAPTER 2

Prem B. Parajuli and Ying Ouyang (2013). Watershed-Scale Hydrological Modeling Methods and Applications, Current Perspectives in Contaminant Hydrology and Water Resources Sustainability, Dr. Paul Bradley (Ed.), ISBN: 978-953-51-1046-0, InTech, DOI: 10.5772/53596.

CHAPTER 3

Yingying Meng, Huixiao Wang, Jiangang Chen, and Shuhan Zhang, "Modelling Hydrology of a Single Bioretention System with HYDRUS-1D," The Scientific World Journal, vol. 2014, Article ID 521047, 10 pages, 2014. doi:10.1155/2014/521047

CHAPTER 4

Dong Jiang, Jianhua Wang, Yaohuan Huang, Kang Zhou, Xiangyi Ding, and Jingying Fu, "The Review of GRACE Data Applications in Terrestrial Hydrology Monitoring," Advances in Meteorology, vol. 2014, Article ID 725131, 9 pages, 2014. doi:10.1155/2014/725131

CHAPTER 5

Saeid Eslamian, Kristin L. Gilroy and Richard H. McCuen (2011). Climate Change Detection and Modeling in Hydrology, Climate Change - Research and Technology for Adaptation and Mitigation, Dr Juan Blanco (Ed.), ISBN: 978-953-307-621-8, InTech, DOI: 10.5772/24550.

CHAPTER 6

Ivan N. da Silva, José Ângelo Cagnon and Nilton José Saggioro (2013). Recurrent Neural Network Based Approach for Solving Groundwater Hydrology Problems, Artificial Neural Networks - Architectures and Applications, Prof. Kenji Suzuki (Ed.), ISBN: 978-953-51-0935-8, InTech, DOI: 10.5772/51598.

CHAPTER 7

C. Day and K. Bremer, "Modeling Urban Hydrology: A Comparison of New Urbanist and Traditional Neighborhood Design Surface Runoff," International Journal of Geosciences, Vol. 4 No. 5, 2013, pp. 891-897. doi:10.4236/ijg.2013.45083.

CHAPTER 8

Paul R. Houser, Gabriëlle J.M. De Lannoy and Jeffrey P. Walker (2012). Hydrologic Data Assimilation, Approaches to Managing Disaster - Assessing Hazards, Emergencies and Disaster Impacts, Prof. John Tiefenbacher (Ed.), ISBN: 978-953-51-0294-6, InTech, DOI: 10.5772/31246.

CHAPTER 9

Giuseppe Bombino, Vincenzo Tamburino, Demetrio Antonio Zema and Santo Marcello Zimbone (2011). Hydrological Effects of Different Soil Management Practices in Mediterranean Areas, Soil Erosion Issues in Agriculture, Dr. Danilo Godone (Ed.), ISBN: 978-953-307-435-1, InTech, DOI: 10.5772/24216.

CHAPTER 10

Mercado, V., Bâ, K., Quentin, E., Madrid, F. and Gama, L. (2015) Hydrological Model to Simulate Daily Flow in a Basin with the Help of a GIS. Open Journal of Modern Hydrology, 5, 58-67. doi: 10.4236/ojmh.2015.53006.

CHAPTER 11

Johanna Springer, Ralf Ludwig, and Stefan W. Kienzle (2015). Impacts of forest fires and climate variability on the hydrology of an alpine medium sized catchment in the Canadian Rocky Mountains, Hydrology 2015, 2(1), 23-47; doi:10.3390/hydrology2010023

CHAPTER 12

Christian Massari, Luca Brocca, Luca Ciabatta, Tommaso Moramarco, Simone Gabellani, Clement Albergel, Patricia De Rosnay, Silvia Puca, and Wolfgang Wagner (2015). The use of H-SAF soil moisture products for operational hydrology: flood modelling over Italy, Hydrology 2015, 2(1), 2-22; doi:10.3390/hydrology2010002

INDEX